Young People in Digital Environments

Young People in Digital Environments

Agency, Opportunities and Risks

Edited by

Terhi-Anna Wilska

Professor of Sociology, Department of Social Sciences and Philosophy, University of Jyväskylä, Finland

Jussi Nyrhinen

Senior Researcher, Department of Social Sciences and Philosophy, University of Jyväskylä and University Lecturer, Department of Social Sciences, Lappeenranta-Lahti University of Technology, Finland

EE Edward Elgar PUBLISHING

Cheltenham, UK • Northampton, MA, USA

Published by
Edward Elgar Publishing Limited
The Lypiatts
15 Lansdown Road
Cheltenham
Glos GL50 2JA
UK

Edward Elgar Publishing, Inc.
William Pratt House
9 Dewey Court
Northampton
Massachusetts 01060
USA

Authorised representative in the EU for GPSR queries only: Easy Access System Europe – Mustamäe tee 50, 10621 Tallinn, Estonia, gpsr.requests@easproject.com

A catalogue record for this book
is available from the British Library

Library of Congress Control Number: 2025937372

This book is available electronically in the **Elgar**online
Sociology, Social Policy and Education subject collection
https://doi.org/10.4337/9781035329250

MIX
Paper | Supporting responsible forestry
FSC® C013056
FSC www.fsc.org

ISBN 978 1 0353 2924 3 (cased)
ISBN 978 1 0353 2925 0 (eBook)
ISBN 978 1 0353 7042 9 (ePub)

Printed and bound in Great Britain by
TJ Books, Padstow, Cornwall

Contents

Figures

Tables

About the editors

Nyrhinen, Jussi (PhD) is a university lecturer at the Department of Social Sciences, Lappeenranta-Lahti University of Technology, Finland. He also works as a senior researcher at the Department of Social Sciences and Philosophy, University of Jyväskylä, Finland. His research focuses on topics such as omnichannel consumption and problematic online behaviours such as impulsive and compulsive online shopping. Nyrhinen has published his research articles in journals such as *Transport Policy*, *Computers in Human Behavior*, and *The International Review of Retail, Distribution and Consumer Research*. He has also authored chapters for books published by Oxford University Press and Taylor & Francis.

Wilska, Terhi-Anna (PhD) is a professor of sociology at the Department of Social Sciences and Philosophy, University of Jyväskylä, Finland. Her research focuses on consumption and lifestyles, digitalization and ICTs, age and generation, well-being and sustainability. In her current research, she focuses particularly on young people in commercial digital environments, such as social media. She is a co-editor of the *Routledge Handbook on Consumption* (Routledge, 2017) and *Digital Technologies and Generational Identity. ICT Usage Across the Life Course* (Routledge, 2017). She has published a large number of articles in high-ranked journals and contributed to many international edited books published by leading publishers.

Contributors

Arto K. Ahonen (PhD) is a senior researcher at the Finnish Institute for Educational Research at the University of Jyväskylä, Finland. His research focuses on a large variety of educational research, especially recently on the assessment and secondary analysis of PISA data.

Emilie Bossens (MA) is a doctoral researcher at the Meaningful Interactions Laboratory within the Faculty of Social Sciences, KU Leuven. Belgium. Her research focuses on human–computer interaction, online interactions, well-being, mixed-methods, and science communication.

Leen d'Haenens (PhD) is a professor at the Institute for Media Studies within the Faculty of Social Sciences, KU Leuven, Belgium. Her research focuses on the realm of digital media and youth, with a specific emphasis on vulnerabilities.

Verónica Donoso (PhD) is a senior researcher at the Institute for Media Studies, within the Faculty of Social Sciences, KU Leuven, Belgium. Her work focuses on the development and enhancement of intervention programmes related to media literacy and digital skills.

Arne Dulsrud (PhD) is a research professor and research director at Consumer Research Norway, Oslo Metropolitan University, Norway. His research focuses on the transformation of retailing, food consumption, marketing research, consumer markets, consumer choice and grocery retail.

Jodi Dworkin (PhD) is a professor and extension specialist at the Department of Family Social Science, University of Minnesota, USA. Her research focuses on promoting positive family development, parenting adolescents and college students, and the role of technology in these relationships.

Arnd Florack (PhD) is a professor and leader of the Applied Social Psychology and Consumer Research Laboratory, University of Vienna, Austria. His research focuses on how social environments and personality factors influence consumer attention, judgement, self-regulation and behaviour.

Myria Georgiou (PhD) is a professor at the Department of Media and Communications, London School of Economics and Political Science, UK. Her research examines the politics and cultures of migration in the context of intense mediation and urbanization.

Rita Grácio (PhD) is an assistant professor at Lusófona University, Portugal. Her research focuses on cultural and creative industries, audience studies, cultural production, gender studies, and digital culture/digital communication.

Marco Hubert (PhD) is a professor at the Department of Management, Aarhus University, Denmark. His research focuses on topics related to digital marketing, consumer behaviour in digital environments, and information systems research.

Ana Jorge (PhD) is an associate professor at Lusófona University and Senior Researcher at CICANT, Portugal. Her research focuses on audiences, digital culture, children, youth and media, celebrity and influencer culture.

Teresa Koch (MSc) is a doctoral candidate at the Applied Social Psychology and Consumer Research Laboratory, University of Vienna, Austria. Her research focuses on children's digital maturity and positive use of technology, including practical approaches to strengthen students' digital maturity in the school context.

Erlend Kok (MSSc) is a research assistant at Consumer Research Norway, Oslo Metropolitan University, Norway. His research focuses on the vulnerabilities and agency of social movements and individuals.

Franziska Laaber (MSc) is a doctoral candidate at the Applied Social Psychology and Consumer Research Laboratory, University of Vienna, Austria. Her research focuses on young people's digital maturity with the aim of providing evidence-based knowledge on healthy digital media use and its measurement.

Mikko Laamanen (PhD) is a research professor at Consumer Research Norway, Oslo Metropolitan University, Norway. His research focuses on the everyday politics of technology, inclusion, and social change.

Kati Laine (MA Ed., MSc Econ.) is a project researcher at the Finnish Institute for Educational Research at the University of Jyväskylä, Finland. Her research focuses on the relationship between education and working life, student assessment, and financial literacy.

Lijun Li (PhD) is a post-doctoral scholar in the Department of Psychology, Penn State University, USA. Her research focuses on the environmental and genetic impacts on child development.

Vilma Luoma-aho (PhD) is a professor and vice dean at the University of Jyväskylä School of Business and Economics, Finland. Her research focuses on strategic communication of organizations in today's dynamic environment that is increasingly prone to different forms of influence.

Maris Männiste (PhD) is a lecturer in critical data studies at the Institute of Social Studies, University of Tartu, Estonia. Her research focuses on datafication, critical data studies, automated welfare, automated decision-making, and artificial intelligence.

Christopher McMahon (PhD) is a lecturer at the Department of Communication and Media, University of Liverpool, UK. His research focuses on the social, cultural, and ideological significance of videogames, neoliberalism and play, microtransactions in videogames, and football fandom.

Julia Nuckols (MSSc) is a doctoral researcher at the University of Jyväskylä and a project researcher at the University of Turku, Finland. Her research focuses on young people, their life course trajectories, and the challenges within them such as the COVID-19 pandemic.

Francisca Porfírio (MSc) is a communication sciences PhD Student at Lusófona University, Portugal. Her research focuses on the topics of digital parenting and social media.

Mette Ranta (PhD) is a senior specialist at the National Legal Services Authority, Finland. She also works as a university researcher at the Faculty of Educational Sciences, University of Helsinki, Finland. Her research focuses on the life course transition to adulthood, financial capability and financial well-being, and agency.

Eero Rantala (MSSc) works as a researcher and lecturer at the Department of Social Sciences, Lappeenranta-Lahti University of Technology, Finland. Rantala's specialties are statistical and causal modelling in social sciences.

Hanna Reinikainen (PhD) is a post-doctoral researcher at the Centre for Consumer Society Research, University of Helsinki, Finland. Her research focuses on interaction in digital environments and the role of social media influencers in consumer behaviour.

Joyce Serido (PhD) is a professor emerita in the College of Education and Human Development, University of Minnesota, USA. Her research focuses on financial behaviour as a bidirectional and unfolding family process, for example, financial parenting, financial behaviour and economic well-being, and financial behaviour and romantic relationships.

Andra Siibak (PhD) is a professor of media studies, a deputy head of research and development and a programme director of the doctoral curriculum in Media and Communication at the Institute of Social Studies, University of Tartu, Estonia. Her research focuses on opportunities and risks surrounding internet use, cyber risks and harms, and inter-generational family life and communication practices in the era of datafication and platformization.

Gintautas Silinskas (PhD) is an academy research fellow (Academy of Finland) at the Department of Psychology, University of Jyväskylä, Finland. His research focuses on interrelations between parents, teachers, and children's learning, motivational, and behavioural outcomes.

Anu Sirola (PhD) is a university teacher at the Faculty of Social Science, University of Tampere, Finland. Her research focuses on behavioural addictions and other risk behaviours in the context of digitalization from a social psychological perspective.

Dag Slettemeås (MA) is a researcher at Consumption Research Norway, Oslo Metropolitan University, Norway. His research focuses on digital media, economies and consumption, digital literacy, and digital vulnerability.

Niina Sormanen (PhD) is a senior researcher at the Faculty of Social Sciences, University of Turku, Finland. Her research focuses on young people's journalistic media consumption in relation to their expression of civic and political issues, participation, and media literacy in the digital and social media age.

Sonali Srivastava (PhD) is a post-doctoral researcher at the Department of Social Sciences and Philosophy, University of Jyväskylä, Finland. Her research focuses on children as consumers in digital environments, especially children's online privacy.

Xiaoran Sun (PhD) is an assistant professor at the Department of Family Social Science, College of Education and Human Development, University of Minnesota Twin Cities, USA. Her research focuses on using passive sensing to examine adolescents' and parents' smartphone use behaviours and the implications of these behaviours on their relationships and well-being.

Katerina Tsetsura (PhD) is the Gaylord Family Professor of Public Relations and Strategic Communication at the Gaylord College of Journalism and Mass Communication, University of Oklahoma, USA. Her research includes social construction of public relations and understanding and countering disinformation.

Alia Zaki (MA) is a research assistant at the Department of Media and Communications, London School of Economics and Political Science, UK. Her research focuses on conflict resolution and peace-building and on how young refugees and asylum seekers use digital technologies.

Acknowledgements

This book is the result of the enthusiasm of many researchers in several research projects in local and international networks. The core ideas for this book crystallized during a high number of thought-provoking research seminars organized by several research networks, research groups, and projects. Those include the interdisciplinary seminars of the *Digital Life* research group and the seminars and meetings of the projects *#Agents—young people's Agency in Social Media, Building Customer Experience in the Metaverse (METAyCX), DigiConsumers—learning to be consumers in digitalizing societies, CoronaConsumers, Data Analytics and Customer Insight (DAWN)* and *Re-Defining Digital Literacy in Europe (REMEDIS).* As a result of many intriguing discussions, joint article writing, and sessions for planning future research, the idea of this book finally took shape. However, there is usually a long process between an idea and a book. For making the process successful, we wish to express our deep gratitude to all authors who contributed to the book chapters. We also thank the funders that have sponsored us and many of the authors. The Research Council of Finland, the Strategic Research Council established within the Research Council of Finland, Business Finland, and the European Commission are among the main funders. In addition, we wish to thank the independent reviewers for their valuable work, as well as the proofreaders whose contribution is essential in producing good academic text in English. Finally, we would like to express our special gratitude to Beatrice McCartney, Helen Daniels, and Victoria Bromley at Edward Elgar for your support throughout the process.

April, 2025
Terhi-Anna Wilska and Jussi Nyrhinen

1. Digital youth in a digital world – opportunities and constraints: an introduction

Terhi-Anna Wilska and Jussi Nyrhinen

1.1 INTRODUCTION

The digitalization of everyday life has brought forth changes in many mundane practices of individuals, such as communication, media participation, consumption, financial behaviour, work, and leisure. For young people, this change has been particularly drastic. Furthermore, many challenges related to digitalization are interconnected with the larger economic and social transformations today's young people face. How can they keep up with the global turbulence and digital leap that accelerated during the COVID-19 pandemic, followed by the war in Ukraine and the subsequent influx of refugees, information warfare, and economic crisis? How can they maintain their well-being and resilience amidst digitalized and commercialized communication filled with risks to privacy and security? How can they tackle issues related to the expansion of technologies based on artificial intelligence (AI)?

The objective of this book is to address the above-mentioned challenges in a novel way. In existing research, young people are usually presented either as active agents or as objects and targets of digital information influence and commercial communication. There is a large number of studies on young people's use of digital devices and social media, but the research is fragmented and yields controversial results, some emphasizing overly negative and others overly positive aspects of digitization. However, in contemporary digital environments, such as social media, the roles of young people cannot be oversimplified to merely active agents or passive targets and victims. Consequently, a broader and more flexible approach is required. This book contributes to that need by focusing on digitalization in various contexts of young people's lives, such as news and information consumption, AI-based commercial communication, family communication and private data sharing, digital financial literacy and well-being, commercial privacy and vulnerability, platform work, gaming, gambling, AI-based tools in education, digital tools in young immigrants' lives, and dimensions of digital maturity in children and young people. The unifying theme is young people's agency, including risks and opportunities related to it.

Researchers from various disciplines, including sociology, psychology, social psychology, education, media and communication studies, information science, and business studies, hailing from 9 countries (Austria, Belgium, Denmark, Estonia, Finland, Norway, Portugal, the United Kingdom, and the United States), examine the psychosocial, technical, and communicational hurdles faced by young people in diverse spheres of digitalizing societies. Alongside country-specific topics and analyses of global phenomena, numerous chapters also incorporate cross-cultural comparisons. Furthermore, the opportunities and protective factors offered by digital media and the economy during crises are thoroughly analysed and discussed.

1.2 DIGITAL AGENCY OF THE DIGINATIVES

In this book, the research on young people concerns individuals aged approximately between 10 and 30 but also families with smaller children. The age cohorts thus represent the so-called *digital generations*, or *diginatives* – individuals who have been surrounded by digital devices since their early childhood or birth, in contrast to previous generations, frequently referred to as *digital immigrants* (Fortunati et al., 2019; Prensky, 2001; Taipale et al., 2017). In addition, *Generation Z* (born between the late 1990s and early 2010s) and *Generation Alpha* (born after the early 2010s) are the first to grow up with the wide presence of various AI-based technologies (Chan & Lee, 2023; McCrindle, 2021).

Young people were the earliest adopters of information and communication technology (ICT) devices, such as mobile phones, in the 1990s (e.g., Turkle, 1996). Soon, it was obvious that mobile technology had become entangled with economic activity. Researchers observed a connection between young people's consumption patterns and their use of mobile phones, computers, and the internet in the early 2000s (Wilska, 2003; Wilska & Pedrozo, 2007). Additionally, children and young people started spending money on virtual goods within online games, such as Habbo Hotel and World of Warcraft (Lehdonvirta et al., 2009). Within just a couple of decades, social media platforms have become the most important channels for advertising and other marketing communications (Alalwan et al., 2017).

Similar to mobile technology in the 1990s, young people were the earliest adopters of social media in the 2000s and continued to use it at high levels. In the European Union, 84 per cent of young people aged 16–29 used some form of social media in 2022 (Eurostat, 2024). According to the Pew Research Center, in 2023, 93 per cent of US teenagers aged 13–17 used YouTube, and about 60 per cent used Snapchat, TikTok, and Instagram (Anderson et al., 2023). Social media is important for communicating with peers (Gray, 2018) but also in familial communication that has been shifting to social media platforms. This has generated both opportunities and challenges for the everyday practices of families. For instance, sharing information with other family members, particularly children (i.e., 'sharenting') (Blum-Ross & Livingstone, 2020), has raised new issues regarding the privacy of underage family members (Kallioharju et al., 2023). Moreover, digital applications are increasingly used by parents in the surveillance of children, thus compromising children's right to privacy and generally increasing feelings of insecurity and distrust among both parents and children (Sukk & Siibak, 2021).

Researchers in many disciplines are constantly investigating the impact of digital environments on several issues relevant to young people, and both high risks and opportunities have been detected. For instance, studies have found that social media and smartphones relate to teenagers' self-perception, social skills, sleep disturbances, mental health outcomes, insecurity, and life management (Hietajärvi et al., 2022; Salmela-Aro et al., 2017; Tilleczek, 2019). Social media platforms play a key role in the formation of today's young people's personal and social identities (Fujita et al., 2018; Mannerström et al., 2018). Along with the commercialization of social media, financial identity (Bosch et al., 2016) and consumer identity (Wilska et al., 2023) are intertwined with personal and social identities.

However, research has shown that engagement with social media also has substantial benefits. Social media activity can provide social and professional connections, a means for increasing self-esteem, and possibilities for sharing experiences, thereby increasing young people's social capital and strengthening their agency (MacFarlane & Kushner, 2023; Tuominen et al., 2022). Young people demonstrate various forms of agency in private, social, societal, and national contexts. *Individual agency* refers to the capacity to act independently and have the power to make one's own free choices (Spencer & Doull, 2015). Particularly in learning contexts, *co-agency* refers to the interactive, mutually supportive relationships that help young people progress towards their valued goals (Salmela-Aro, 2016).

Young people also promote collective actions on social media platforms, where people with similar desires can easily find each other and coordinate actions to create a desired outcome. This kind of *distributed agency* (Enfield & Kockelman, 2017) is a typical form of young people's agency on social media, where personal influence is possible for anybody. Most young people possess proficient digital abilities, with many of them actively participating and even playing influential roles on digital platforms. On the other hand, young people's familiarity with digital tools often leads to a lack of appreciation for their full capabilities, and they may rely on passive usage rather than actively exploring their potential (Kirschner & De Bruyckere, 2017).

1.3 RISKS AND OPPORTUNITIES IN DIGITAL ENVIRONMENTS

Young people are frequent targets of commercial and non-commercial influence in digital environments. They typically search for information and news from digital platforms that are easy to manipulate (Compton et al., 2021). Therefore, young people's digital literacy (i.e., mechanical and navigational competences in digital environments) needs constant attention. In addition to technical skills, today's digital literacy includes the knowledge and attitudes to use digital media in responsible, critical, and creative ways (Hatlevik et al., 2015) that support one's well-being (Vissenberg et al., 2022). New digital vulnerabilities have emerged as different entities attempt to influence young people's emotions, attitudes, and behaviours, ranging from purchase intentions to information warfare. This makes brands and organizations increasingly responsible for ethical choices and the transparency (i.e., disclosure of sources that might have influenced information) (Tsetsura & Kruckeberg, 2017) of information. Young people are also exposed to the heavy targeting of commercial communication. Social media influencers are particularly impactful in young people's lives, shaping their values, attitudes, and consumption patterns (Djafarova & Rushworth, 2017; Reinikainen et al., 2020). Moreover, AI-driven algorithmic targeting is steadily increasing (Aragoncillo & Orus, 2018). Privacy invasions are among the key problems in commercial digital environments, as young people are heavily exposed to AI-based commercial profiling (Srivastava et al., 2023). In addition, in education, AI-based profiling and evaluation may enhance inequality and hinder future opportunities for vulnerable children and young people (Mascheroni & Siibak, 2021).

In digital marketplaces, today's young people become consumers and economic actors at an early age (Buckingham & Willett, 2022). It is not only consumption and advertising that occur on online platforms but also the financial services of banks, for example, saving, investing, taking loans, and obtaining debit and credit cards. Thus, financial literacy (i.e., knowledge, skills, confidence, and motivation) is necessary to effectively manage money (Lusardi, 2015), but also digital financial literacy is also needed, that is, skills to effectively use digital technologies for financial matters: doing online shopping, using online and mobile banking, using financial services, and avoiding risks and fraud in digital environments (e.g., Koskelainen et al., 2023). Digital financial literacy is increasingly important for responsible consumption and money management, including the competence to protect oneself from financial cybercrimes, such as credit card fraud and identity theft (Andreou & Anyfantaki, 2021; Mohd Padil et al., 2022).

Active engagement in digital environments is also connected to several other issues, such as problematic online behaviours. These include different kinds of addictive behaviours, such as online shopping addiction (Nyrhinen et al., 2023), gambling, and gaming (Sirola et al., 2023). Furthermore, problems have arisen from the platformization of work, for example, in professional tasks (e.g., graphic design) or application-based transport and delivery services. Platform work is independent and flexible but may also be financially more insecure than traditional employment, particularly for young people and immigrants (van Doorn & Vijay, 2024).

1.4 THE STRUCTURE OF THE BOOK

This book presents a diverse array of studies based on recent quantitative and qualitative international data collected from young people, young adults and their families, as well as research reviews and conceptual analyses. The book is divided into three parts. The first part, 'Young people's agency in public and private digital communication', specifically focuses on how young people communicate and interpret information obtained from digital environments – also when targeted with AI-based influencing. Furthermore, the first part sheds light on digital communication practices in families.

In Chapter 2, Niina Sormanen and Hanna Reinikainen write a research review on the impact of diverse social media sources on the way young people consume news and other information. The chapter also explores how the unique features of digital and social media platforms have shaped the news and information needs and gratifications of today's youth. In Chapter 3, Jussi Nyrhinen and Eero Rantala investigate the impact of AI on young people's agency within commercial digital contexts. Utilizing technology analysis, the chapter reveals that young consumers inadvertently influence AI outcomes by leaving data traces through their digital behaviours, thereby shaping consumer behaviour and agency in digital environments. In Chapter 4, Katerina Tsetsura, Vilma Luoma-aho, and Eero Rantala present a novel empirical analysis of young people's attitudes towards the quality of social media information in Finland and the United Kingdom. The research detects notable differences in attitudes towards the non-transparency of information, but reveals that careful and critical online media consumers form the largest group of respondents.

Jodi Dworkin and Xiaoran Sun analyse patterns of digital communication between young people and their parents in Chapter 5. Based on a survey and data collected from young people's smartphones, the research shows that parents and young people report frequent and variable communication both digitally and in person. Digital communication, however, did not replace face-to-face communication. In Chapter 6, Ana Jorge, Francisca Porfírio, and Rita Grácio report the results of an ethnographic study that explores families in the processes of datafication and digitalization through the influence of school and peers. The study highlights differences in how parents and children perceive privacy and engage with digital media, for example, regarding 'sharenting' (i.e., sharing information on family members beyond the home environment).

In the second part, 'Young people as digital consumers and financial actors', the focus turns to young people's agency and position within economic behaviour in commercial digital environments. The importance of digital environments in economic behaviour, such as online purchases and the utilization of digital financial services, has notably increased, particularly along with the COVID-19 restrictions. Thus, the chapters examine young people's digital financial literacy, encompassing their financial knowledge and financial identity in digital settings – also during the pandemic.

In Chapter 7, Gintautas Silinskas, Kati Laine, and Arto K. Ahonen analyse how the active use of ICT is connected to teenagers' financial literacy. Based on data derived from the Programme for International Student Assessment (PISA) of eight European countries among 15-year-old adolescents, the research shows that financial literacy is associated with ICT competence and financial confidence and autonomy. Mette Ranta, Lijun Li, and Joyce Serido explore the role of like-minded (homophilous) social media groups in shaping young adults' financial identity and well-being during the COVID-19 pandemic in Chapter 8. Using survey data, the study reveals that young adults experience conditional benefits related to social media use, depending on factors such as socioeconomic position and peer support.

In Chapter 9, Mikko Laamanen, Erlend Kok, Arne Dulsrud, and Dag Slettemeås investigate how young adults perceive their digital empowerment and vulnerability in the context of personalized online advertising on social media platforms. With a two-part qualitative study, the perceptions are studied from the viewpoints of the media and advertising industry, interest organizations and market regulators. In Chapter 10, Sonali Srivastava examines the factors that prevent young people from protecting their commercial privacy in digital environments. Focus group discussions reveal that young people lack awareness or understanding of privacy practices. Challenges also arise from how privacy protection mechanisms are implemented in digital spaces.

Chapter 11 highlights microtransactions in games. Christopher McMahon critically analyses how microtransactions in video games such as the *FIFA* (now *EA Sports FC*) series and *Fortnite* are intended to monetize the effects of neoliberal ideology operating within a video game. This chapter describes how young people exercise agency in their everyday lives within these digital play spaces. In Chapter 12, Julia Nuckols explores the rise of digital gig work (i.e., application-based platform services such as food delivery or ridesharing, for example, Wolt, DoorDash, and Uber), weighing its strengths and weaknesses, juxtaposed with the uncertainties typical to the younger age.

The third part of the book, 'Risks and opportunities for young people in digital environments', illuminates the challenges and opportunities that digital environments present to young people, such as gambling, problems with AI-based grading and evaluation in education, affordances of digital tools for the agency of young refugees, and, finally, new evidence-based conceptualizations to understand and support young people's digital agency.

Chapter 13 synthesizes research evidence on internet-based problematic consumption behaviours among young adults during the COVID-19 pandemic, namely excessive online gambling, gaming, and shopping. Anu Sirola analyses the risks to the emotional, social, and financial well-being of these behaviours to young people who are typical consumers of these digital services, which have addictive potential. In Chapter 14, Maris Männiste and Andra Siibak examine bias, (un)fairness, and discrimination in the grading of students resulting from the use of AI-based technologies in the education sector, as well as the use of online proctoring software to combat academic dishonesty. The chapter showcases that despite such technologies appearing seemingly neutral and fair, their automatic biases and built-in fairness can increase inequality.

Chapter 15 presents a field study conducted in European cities. Myria Georgiou, Leen d'Haenens, Alia Zaki, Verónica Donoso, and Emilie Bossens illustrate how teenage refugees mobilize individual and collective agency with the support or disadvantage of online tools, to manage daily and ontological needs for connection, reassurance, and stability when living under conditions of vulnerability. In Chapter 16, Franziska Laaber, Teresa Koch, Arnd Florack, and Marco Hubert provide a comprehensive novel theoretical conceptualization of digital maturity and its measurement. With empirical evidence, this chapter illustrates the factors that shape digital maturity and highlights how these elements explain the differences in digital competence among young people.

In the final chapter, the main themes of the book are summarized and discussed, considering practical implications and future research. Based on the multiple contexts of young people's lives, it is concluded that today's digitalized societies provide young people with significant opportunities but also present serious risks. With the rapid development of AI-based technologies and with the increasing commercialization of digital media, it is crucial to anticipate future digital vulnerabilities that could undermine young people's agency and well-being and to find tools to build their resilience and support their agency. It is imperative that young people acquire enhanced media literacy, digital literacy, and financial literacy to fully capitalize on the opportunities afforded by digital environments. Moreover, since AI is changing the digital landscape, traditional media literacy and consumer education programmes may become outdated. We conclude that for the current and future AI generations, better AI literacy is needed to support young people's agency in everyday life, communication, consumption, education, and work in the future. This book presents various ways of achieving that goal.

REFERENCES

Alalwan, A. A., Rana, N. P., Dwivedi, Y. K., & Algharabat, R. (2017). Social media in marketing: A review and analysis of the existing literature. *Telematics and Informatics*, 34(7), 1177–1190. https://doi.org/10.1016/j.tele.2017.05.008.

Anderson, M., Faverio, M., & Gottfried, J. (2023). *Teens, Social Media and Technology*. Pew Research Center. https://www.pewresearch.org/wp-content/uploads/sites/20/2023/12/PI_2023.12.11-Teens-Social-Media-Tech_FINAL.pdf.

Andreou, P. C., & Anyfantaki, S. (2021). Financial literacy and its influence on internet banking behavior. *European Management Journal*, 39(5), 658–674. https://doi.org/10.1016/j.emj.2020.12.001.

Aragoncillo, L., & Orus, C. (2018). Impulse buying behaviour: An online-offline comparative and the impact of social media. *Spanish Journal of Marketing – ESIC*, 22(1), 42–62. https://doi.org/10.1108/sjme-03-2018-007.

Blum-Ross, A., & Livingstone, S. (2020). 'Sharenting,' parent blogging, and the boundaries of the digital self. In Thumin, N. (Ed.), *Self-(re) Presentation Now*. Abingdon: Routledge, pp. 70–85.

Bosch, L. A., Serido, J., Card, N. A., Shim, S., & Barber, B. (2016). Predictors of financial identity development in emerging adulthood. *Emerging Adulthood*, 4(6), 417–426. https://doi.org/10.1177/216769681663184.

Buckingham, D., & Willett, R. (2022). Constructing children as consumers. In *The Routledge International Handbook of Children, Adolescents, and Media*. Abingdon: Routledge, pp. 49–56.

Chan, C. K. Y., & Lee, K. K. (2023). The AI generation gap: Are Gen Z students more interested in adopting generative AI such as ChatGPT in teaching and learning than their Gen X and millennial generation teachers? *Smart Learning Environments*, 10(1), 60.

Compton, J., van der Linden, S., Cook, J., & Basol, M. (2021). Inoculation theory in the post-truth era: Extant findings and new frontiers for contested science, misinformation, and conspiracy theories. *Social and Personality Psychology Compass*, 15(6), e12602. https://doi.org/10.1111/spc3.12602.

Djafarova, E., & Rushworth, C. (2017). Exploring the credibility of online celebrities' Instagram profiles in influencing the purchase decisions of young female users. *Computers in Human Behavior*, 68, 1–7. https://doi.org/10.1016/j.chb.2016.11.009.

Enfield, N. J., & Kockelman, P. (Eds). (2017). *Foundations of Human Interaction: Distributed Agency*. Oxford: Oxford University Press.

Eurostat. (2024). Young people-digital world. https://ec.europa.eu/eurostat/statistics-explained/index.php?title=Young_people_-_digital_world Data extracted May, 2024.

Fortunati, L., Taipale, S., & de Luca, F. (2019). Digital generations, but not as we know them. *Convergence*, 25(1), 95–112. https://doi.org/10.1177/1354856517692309.

Fujita, M., Harrigan, P., & Soutar, G. N. (2018). Capturing and co-creating student experiences in social media: A social identity theory perspective. *The Journal of Marketing Theory and Practice*, 26(1–2), 55–71. https://doi.org/10.1080/10696679.2017.1389245.

Gray, L. (2018). Exploring how and why young people use social networking sites. *Educational Psychology in Practice*, 34(2), 175–194. https://doi.org/10.1080/02667363.2018.1425829.

Hatlevik, O. E., Guðmundsdóttir, G. B., & Loi, M. (2015). Examining factors predicting students' digital competence. *Journal of Information Technology Education: Research*, 14, 123–137.

Hietajärvi, L., Maksniemi, E., & Salmela-Aro, K. (2022). Digital engagement and academic functioning. *European Psychologist, 27*(2), 102–115. https://doi.org/10.1027/1016-9040/a000480.

Kallioharju, M., Wilska, T. A., & Vänskä, A. (2023). Mothers' self-representations and representations of childhood on social media. *Young Consumers*, 24(4), 485–499. https://doi.org/10.1108/YC-06-2022-1541.

Kirschner, P. A., & De Bruyckere, P. (2017). The myths of the digital native and the multitasker. *Teaching and Teacher Education*, 67, 135–142.

Koskelainen, T., Kalmi, P., Scornavacca, E., & Vartiainen, T. (2023). Financial literacy in the digital age – A research agenda. *Journal of Consumer Affairs*, 57(1), 507–528.

Lehdonvirta, V., Wilska T.-A., & Johnson, M. (2009). Virtual consumerism. *Information Communication & Society*, 12(7), 1059–1079. https://doi.org/10.1080/13691180802587813.

Lusardi, A. (2015). Financial literacy skills for the 21st century: Evidence from PISA. *Journal of Consumer Affairs*, 49(3), 639–659.

MacFarlane, B., & Kushner, J. (2023). Social media and young adults. In Batra, K. & Sharma, M. (Eds), *Effective Use of Social Media in Public Health*. Cambridge, MA: Academic Press, pp. 45–63. https://doi.org/10.1016/B978-0-323-95630–7.00010-X.

Mannerström, R., Hietajärvi, L., Muotka, J., & Salmela-Aro, K. (2018). Identity profiles and digital engagement among Finnish high school students. *Cyberpsychology: Journal of Psychosocial Research on Cyberspace*, 12(1), Article 2. https://doi.org/10.5817/cp2018-1-2.

Mascheroni, G., & Siibak, A. (2021). *Datafied Childhoods. Data Practices and Imaginaries in Children's Lives*. Oxford: Peter Lang.

McCrindle, M. (2021). *Generation Alpha*. London: Hachette.

Mohd Padil, H., Kasim, E. S., Muda, S., Ismail, N., & Md Zin, N. (2022). Financial literacy and awareness of investment scams among university students. *Journal of Financial Crime*, 29(1), 355–367. https://doi.org/10.1108/JFC-01-2021-0012.

Nyrhinen, J., Lonka, K., Sirola, A., Ranta, M., & Wilska, T. A. (2023). Young adults' online shopping addiction: The role of self-regulation and smartphone use. *International Journal of Consumer Studies*, 47(5), 1871–1884. https://doi.org/10.1111/ijcs.12961.

Prensky, M. (2001). Digital natives, digital immigrants part 2: Do they really think differently? *On the Horizon*, 9(6), 1–6. https://doi.org/10.1108/10748120110424843.

Reinikainen, H., Munnukka, J., Maity, D., & Luoma-aho V. (2020). 'You really are a great big sister'– Parasocial relationships, credibility, and the moderating role of audience comments in influencer marketing. *Journal of Marketing Management*, 36(3–4), 279–298. https://doi.org/10.1080/0267257x.2019.1708781.

Salmela-Aro, K. (2016). *Co-agency*. Working papers. Paris: OECD.

Salmela-Aro, K., Upadyaya, A., Hakkarainen, K., Lonka, K., & Alho, K. (2017). The dark side of internet use: Two longitudinal studies of excessive internet use, depressive symptoms, school burnout and engagement among Finnish early and late adolescents. *Journal of Youth and Adolescence*, 46(2), 343–357. https://doi.org/10.1007/s10964-016-0494-2.

Sirola, A., Nyrhinen, J., Nuckols, J., & Wilska, T. A. (2023). Loot box purchasing and indebtedness: The role of psychosocial factors and problem gambling. *Addictive Behaviors Reports*, 18, 100516. https://doi.org/10.1016/j.abrep.2023.100516.

Spencer, G., & Doull, M. (2015). Examining concepts of power and agency in research with young people. *Journal of Youth Studies*, 18(7), 900–913. https://doi.org/10.1080/13676261.2014.1001827.

Srivastava, S., Wilska, T. A., & Nyrhinen, J. (2023). Children as social actors negotiating their privacy in the digital commercial context. *Childhood*, 30(3), 235–252. https://doi.org/10.1177/09075682231186486.

Sukk, M., & Siibak, A. (2021). Caring dataveillance and the construction of 'good parenting': Estonian parents' and pre-teens' reflections on the use of tracking technologies. *Communications*, 46(3), 446–467. https://doi.org/10.1515/commun-2021-0045.

Taipale, S., Wilska, T. A., & Gilleard, C. J. (Eds). (2017). *Digital Technologies and Generational Identity*. Abingdon: Routledge.

Tilleczek, K. (2019). Young lives in the digital age. In Tilleczek, K. C. & Campbell, V. M. (Eds), *Youth in the Digital Age. Paradox, Promise, Predicament*. Abingdon: Routledge, pp. 1–11.

Tsetsura, K., & Kruckeberg, D. (2017). *Transparency, Public Relations and the Mass Media: Combating the Hidden Influences in News Coverage Worldwide*. New York: Taylor and Francis.

Tuominen, J., Rantala, E., Reinikainen, H., Luoma-aho, V., & Wilska, T. A. (2022). The brighter side of materialism: Managing impressions on social media for higher social capital. *Poetics*, 92, 101651. https://doi.org/10.1016/j.poetic.2022.101651.

Turkle, S. (1996). *Life on the Screen: Identity in the Age of the Internet*. London: Weidenfeld & Nicolson.

van Doorn, N., & Vijay, D. (2024). Gig work as migrant work: The platformization of migration infrastructure. *Environment and Planning A: Economy and Space*, 56(4), 1129–1149. https://doi.org/10.1177/0308518X211065049.

Vissenberg, J., d'Haenens, L., & Livingstone, S. (2022). Digital literacy and online resilience as facilitators of young people's well-being? *European Psychologist*, 27(2), 76–85. https://doi.org/10.1027/1016-9040/a000478.

Wilska, T. A. (2003). Mobile phone use as part of young people's consumption styles. *Journal of Consumer Policy*, 26(4), 441–463. https://doi.org/10.1023/A:1026331016172.

Wilska, T. A., & Pedrozo, S. (2007). New technology and young people's consumer identitites: A comparative study between Finland and Brazil. *Young*, 15(4), 343–368. https://doi.org/10.1177/110330880701500404.

Wilska, T. A., Holkkola, M., & Tuominen, J. (2023). The role of social media in the creation of young people's consumer identities. *Sage Open*, *13*(2), https://doi.org/10.1177/21582440231177030

PART I

Young people's agency in public and private digital communication

2. Young people as audiences, consumers and participants in digital news and information environments

Niina Sormanen and Hanna Reinikainen

2.1 INTRODUCTION

The information and media landscape has been transformed due to digital and social media development, offering people multiple traditional and new journalistic media outlets and platforms from which to choose their daily digest of news and information. Young people, in particular, are growing up in a media landscape that is mostly digital and social, and their media use and motivations differ from those of earlier generations; hence, the concept of a 'new media generation' (Schanke Sundet & Lüders, 2023). The new media landscape provides young people with more opportunities for news selection and curation, along with participation with media providers, content, and other media users, like never before. Young people are still losing touch with traditional journalistic media, such as newspapers and magazines, and the reasons behind their consumption and non-consumption are increasingly being researched and explored by media organisations that aim to produce content that would pique the interest of future news consumers. According to the Reuters Institute report (Newman et al., 2023), the global news industry has struggled through the first waves of the move to a digital, mobile, and platform-dominated media environment, but it still faces a fundamental challenge to capture and keep the attention of young generations who have grown up with and rely mostly on social media platforms.

Moreover, in the current digital age, new types of social media content producers often draw the attention of young people: social media influencers (SMIs). SMIs have been defined as individuals who have gained a significant following by regularly posting content on social media platforms (e.g., Instagram, YouTube, TikTok, and blogs), through interaction and relationship building with their following, and by practising personal branding (Abidin, 2016; Enke & Borchers, 2019). SMIs typically have the ability to evoke experiences of parasocial interaction and parasocial relationships (i.e., illusory, one-sided relationships; see Horton & Wohl, 1956) among their audiences and to capitalise on these relationships and their personal brands through commercial collaborations with brands and organisations (Reinikainen et al., 2020). SMIs can have a significant influence on young people's opinions, attitudes, and behaviours in the realm of consumer behaviour but also in civic and political issues (Suuronen et al., 2022).

This chapter's research review combines research findings and theories that help to understand why young people choose to consume information and news from different sources, in specific forms, and with varied engagement levels in the current digital and social media age, and the consequences of young people's consumption patterns on them, society, and journalism's future. Young people's media and information consumption is an important issue because young people are future consumers of news and other journalistic media. The stature of legitimate media and news and their use for acquiring verified information have consequences for journalism's survival and democracy at large (Nielsen, 2016), especially as the connection between young people's news exposure and participation in civic and political affairs has been revealed in many studies (e.g., Edgerly et al., 2018; Gil de Zúñiga et al., 2017). It is also important to understand the roles of central youth influencers, such as SMIs, in young people's news and information consumption in the modern media landscape.

Research on media audiences has been conducted since the beginning of the twentieth century when audiences were mostly seen as passive receivers of mass media agendas and content. Digitalisation has significantly changed audiences' relationships with the media because audiences gain access to different news and media content through a variety of media, platforms, channels, and sources, which have become part of people's everyday lives (Lunt & Vicente-Mariño, 2015). Audiences have also become more active in engaging and participating with the media, their outputs, other users, and journalists, and they can influence media content and strategies, such as through online feedback, likes, recommendations, and re-shares.

One theory that can be utilised to understand and examine media audiences is the uses and gratifications theory (U&G theory), a framework established in the 1970s by Katz et al. (1973–1974). The U&G theory seeks to explain how individuals use different media, from mass media and newspapers to the internet and social media, to gratify their needs, discover their underlying motives (innate needs) for media use, and identify the positive and negative consequences of individual media use (Huang, 2009). Scholars have determined that many of the uses and gratifications of media use are the same for traditional 'old' (e.g., print newspapers) and 'new' (e.g., online newspapers) journalistic media and also for social media. More importantly, they have also found that the new affordances of today's digital media technology have significantly shaped audiences' media uses, needs, and gratifications (Sundar & Limperos, 2013). In this process, digital and social media have also blurred the lines between journalistic and other information in the digital age because the producers of legacy media and other social media content (e.g., SMIs) are appealing to somewhat similar gratifications of young people's information needs. The consequent potential platform, content, and source confusion can make it difficult, for instance, to identify mis- and disinformation. The MAIN Model (Sundar, 2008; Sundar & Limperos, 2013) is a further elaboration of the U&G theory, concentrating on how specific technological affordances (modality, agency, interactivity, and navigability) produce psychological effects on young people and influence media consumption gratifications.

This book chapter first examines young people's news consumption trends in the current digital and social media era, introducing varied news user and non-user groups, their antecedents, and their noted consequences. The second part focuses more closely on how digital affordances have specifically shaped journalistic content and young people's consumption needs and gratifications, as well as the roles played by actors such as SMIs in the online news and information consumption and gratification process. Hence, the aim is to show how social media U&Gs are intertwined with online news U&Gs and the consequences of this for youth information consumption. The last part of this chapter describes how young people participate in and engage with different forms of journalism and how different media and other information producers may create a blurred and confusing media environment for young people. The Discussion section draws conclusions and further aspects to consider and research. This chapter is not a systematic review of all research on news and social media consumption by young people but a current reflection of how news media, other sources of information, and influencers' content are consumed by young people today because digital and social media platforms greatly affect information consumption and participation.

2.2 LITERATURE REVIEW

2.2.1 Young People's News Consumption in the Digital and Social Media Age

International studies have found that young people still mainly trust established news institutions but are drawn to online and social media news content that provides them with youth-relevant information mostly in short and visual formats, free of charge, when they want it, on social media platforms the young are on, and with an opportunity to interact and participate (e.g., Galan et al., 2018; Huang, 2009; Newman et al., 2023). Older generations continue to prefer news in text form, but as video and audio content have become a part of people's everyday lives through digital and social media platforms, young audiences in particular increasingly expect similar content from news media (Newman et al., 2023; Sundar & Limperos, 2013).

Research on young people's news consumption has been conducted, for instance, through an analysis of news repertoires, which means grouping users depending on their favoured news media use and frequency (e.g., Geers, 2020). Based on studies around the world, common news repertoires are identified as groups of 'minimalists' or 'news avoiders' (usually the largest group), who access news infrequently or avoid news altogether; 'traditionalists', who favour print newspapers and television for news; 'online or new media news consumers', who use mainly the internet, mobile apps, and social media for news; and 'omnivores', who frequently use news from various sources, including social media, and are usually the smallest group (e.g., Lee & Yang, 2014; Vozab, 2019). According to the very few news repertoire studies that have been conducted among young people in the USA and the Netherlands, most of the young people belonging to the group of 'news avoiders' or 'minimalists' are uninterested in news from any source. Of the aforementioned groups, the second largest has been identified as traditionalists; the third, social media news users; and the smallest, news omnivores (Geers, 2020; Edgerly et al., 2018).

One youth survey among 15- to 19-year-olds (Sormanen et al., 2022a) revisited the notion of youth news repertoires (Geers, 2020; Edgerly et al., 2018) and found three youth news consumption groups from Finland somewhat differing from previous studies. The study identified a group of 'moderate digital traditionalists', who quite often consume news primarily through online versions of newspapers, as the largest group. 'Minimalist social media stumblers' were the second largest group, and frequent news omnivores consisted of a rather good proportion of high news followers through many sources, from digital and print to social media. The survey results indicate that the young Finnish audience belongs to a specific suggested Nordic news media system (spanning Denmark, Finland, Norway, and Sweden), which emphasises a preference for newspapers' online editions as the main sources of news instead of social media or search engines, and has a high level of trust in news and a willingness to pay for online news (Schrøder et al., 2020). The Reuters Institute news report (Newman et al., 2023) supports this notion because people in Nordic countries have been found to be highly willing to pay for online news and young adults in Finland and Sweden are even more willing to pay for digital news than older age groups. In addition, although global trust in news has declined significantly, trust in news is still high in Nordic countries, such as Denmark and Norway, and Finland has been evaluated year after year as the country with the highest level of trust in news media (Newman et al., 2023).

The Finnish youth survey study (Sormanen et al., 2022a) also found a significant connection between the use of online news sites and the use of social media platforms to find news among the group of minimalist social media stumblers. This connection means that young people in this low-news-consumption group mostly accidentally come across news on social media platforms and follow them to online news sites, while the other groups also actively search for news. This phenomenon is tied to the digital-era process, where print and digital news are most often purposefully sought while social media includes news links mainly appearing on users' feeds because their social networks share the news or algorithms suggest the news to users depending on their earlier behaviour or personal data (e.g., Boczkowski et al., 2018). Young people in different countries in particular have been found to 'incidentally' notice and access news via social media platforms. 'Incidental' news consumption is connected with the behaviour of reading the news quickly while surfing online, which leads to risks such as low levels of societal knowledge and receiving unchecked information or disinformation (Boczkowski et al., 2018; Gil de Zúñiga et al., 2017). Still, even incidental news consumption through social media can be beneficial, especially for those less attracted by news, because young people can be more interested in news shared by their peers and close networks (Sormanen et al., 2022a).

Many studies have also examined the reasons behind young people's news consumption. Previous studies have connected the following factors to young people's more traditional or frequent and varied news consumption: political knowledge and interests; older age; higher education and socioeconomic status; parents' higher education level and news consumption habits; young people's general information-seeking eagerness; and higher trust in traditional media (e.g., Galan et al., 2018; Geers, 2020; Edgerly et al., 2018; Sormanen et al., 2022a). Research on UK and US youngsters has identified the following central reasons behind young people's low news consumption: time constraints due to school, job, or leisure activities; non-relevance of news topics for young people; the traditional newspaper format being too long and effortful and not where young people spend their time (i.e., social media); parents' non-exposure of their children to news or non-encouragement of their children's news consumption; the news format not fitting the platform and affordances; and youth being overwhelmed by the amount of especially negative news from various sources (Galan et al., 2018; Huang, 2009).

News avoidance is a growing phenomenon in the current digital and social media age. Specific reasons for actively avoiding the news include the heaviness and negativity of news (i.e., repeated stressful news topics), mistrust in news, perceiving news as non-relevant for oneself, low knowledge of the news system (e.g., low news literacy skills), and news and information overload, which causes anxiety and stress (Skovsgaard & Andersen, 2020; Villi et al.,

2021). There are still country- and culture-specific differences found in how news consumption is perceived, for example, as a civic duty (e.g., in Israel) or how politically polarised or draining the context is (e.g., in the USA), which influences news avoidance (Villi et al., 2021). Young and less educated people have been identified as the likeliest to avoid news entirely or occasionally (Newman et al., 2023), and emotional self-preservation from negative news is most common among young people (Villi et al., 2021). After this broader review of news consumption and non-consumption among young people, the following section focuses on the technological affordances that guide young people's uses of news and information online and the gratifications they obtain from their consumption of these.

2.2.2 Young People's Media and Information Needs and Gratifications

The traditional U&G approach assumes that people have innate needs that drive them to select to use different types of media, and that gratifications come when the needs are met or satisfied by certain types of media sources that match people's expectations (e.g., Katz et al., 1973–1974; Sundar & Limperos, 2013). Hence, the approach can be used to examine the audience needs of specific types of media, from traditional to new, and from journalistic media to social media and websites.

The gratifications of print newspaper use in the 1940s were identified as surveillance, interpreting affairs, social prestige, and escape (Sundar & Limperos, 2013), and online newspaper gratifications of the digital age were identified as information seeking, pastime, entertainment and socialisation (Yoo, 2011). Consequently, the gratifications of older and newer news forms can be surprisingly similar. Still, the traditional U&G measures are rooted primarily in the social and psychological origins of needs, while the new media age has raised questions as to whether the traditional U&G approach can capture the gratifications from newer media usage, especially through the affordances provided by media technology. Sundar and Limperos (2013) suggest that the U&Gs, for instance, of news use are highly affected by the affordances provided by digital technologies and platforms, making them more nuanced: news websites allow users to browse news much like print newspapers but also allow clicking links to other content, viewing the news in video form, having conversations with other users and journalists, and sharing the news through other platforms. Hence, the three main divisions of traditional gratification have changed due to digital developments: 'content gratification' comes from a variety of formats (e.g., videos); 'process gratification' comes from the expectation of navigable, interactive elements; and 'social gratification' comes from more varied and timely access to other users and discussions (Sundar & Limperos, 2013).

Sundar (2008) and Sundar and Limperos (2013) also identify four specific classes of technological affordances in digital media: modality, agency, interactivity, and navigability (the MAIN Model), which are found to have significant psychological consequences. According to the MAIN Model, *modality-based gratifications* come from how media content is presented to users (e.g., in audio, picture, or text format), because the formats appeal to different aspects of the human perceptual system. Textual content demands more cognitive effort, audiovisual content can be distracting, and visual content can be more trusted because of a 'realism heuristic', indicating that if something is photographed or shown in video form (e.g., in news), it must be more real than text descriptions. *Agency-based gratifications* come from the way digital and social media allow all users to be agents, sources, and gatekeepers of information, while the role of information and agenda gatekeeping has historically been the domain of a privileged few. In the current digital age, people also want to know what other people and peers consider worthy and trustworthy. By viewing other people's favoured or shared content, people can make decisions, for instance, on whether to read the news or follow an SMI. *Interactivity-based gratifications* come from allowing users to interact with and through the medium. Users expect greater levels of activity and a participatory nature (human interaction) from their media experiences, want the media to be responsive to their actions (e.g., controlling news presentation), and expect more information through hyperlinks and flow in their media experience (e.g., on news websites). Finally, *navigability-based gratifications* come from allowing users to move through the medium. People expect digital spaces to be easily navigated through links and browsing.

Among American youth, Huang (2009) identified the same technological affordances as elements that young people look for in the news in the current digital era. Accordingly, young people often prefer short stories and multimedia visual presentations to bring them not only instant satiation but also the possibility of rich media in-depth reading opportunities. They like to control what news to access and how to access it, as well as to have opportunities for easy content contribution and accessibility through various devices. In addition, young people want not only to be passive news receivers but also to contribute content.

All in all, young people's gratifications from news use still come largely from staying informed about societal events and phenomena (i.e., information seeking), in addition to gratifications such as passing time, entertainment, and socialisation (e.g., Galan et al., 2018; Huang, 2009; Newman et al., 2023; Yoo, 2011). Still, technological considerations, such as the user-friendliness of news platforms and their interactive affordance, can have a significant effect on young people's news consumption (Villi et al., 2021; Sundar & Limperos, 2013). Because young people favour digital and social media affordances that gratify their needs, news media organisations use the same technologies providing interactivity as other digital multimedia (blogs, podcasts, and social media platforms) and have accessed social media platforms with their own pages and news content dissemination.

However, news media are not the only information providers in the digital and social media landscape of the young. Today, young people may get their news even more often from SMIs or other celebrities instead of journalists (Newman et al., 2023). The motivations and gratifications of online news use are surprisingly similar to those of social media use: interpersonal utility (e.g., socialisation), passing time, information seeking, convenience, and entertainment (see Sundar & Limperos, 2013). Hence, general information seeking and news as information sources become intertwined in social media use through the similar motivations of use and the satisfaction they provide.

Young people are typically considered the main audience segment for SMIs (Enke & Borchers, 2019; Croes & Bartels, 2021), and a common notion among media and brand executives is that young people these days are difficult to reach through legacy media and that SMIs offer a clever way to address young people (Reinikainen et al., 2022; Schanke Sundet & Lüders, 2023). However, it appears that the notion of young people being isolated in their SMI bubbles, far away from epistemic and democratic institutions, is at least partly misleading; the consumption of legacy media and even voting appear to be connected to following SMIs (Pöyry & Reinikainen, 2024; Martin & Sharma, 2022), and information seeking and learning about new things are important motivations for following SMIs (Croes & Bartels, 2021). This suggests that SMIs have, in fact, become a regular part of the modern media landscape for young people and that engaging with SMIs has become an important way for young people to stay up to date about current issues (Pöyry & Reinikainen, 2024) and even consume content by legacy media (Martin & Sharma, 2022).

The enchantment of SMIs typically relies on their ability to create intimate and trusting relationships with their followers. SMIs are often considered peers, old friends, or even family members (Abidin, 2016; Reinikainen et al., 2020) whom followers turn to in search of companionship and solace (Croes & Bartels, 2021). At the same time, followers' trusting relationships with SMIs have been considered possibly risky when it comes to the spread of mis- and disinformation. Hasell and Chinn (2023), for example, discovered that following so-called lifestyle influencers appears to be positively associated with 'anti-intellectualism'. Thus, the followers of lifestyle influencers might be prone to, for example, following health advice given by influencers that is not recommended by authorities. During the coronavirus 2019 pandemic, SMIs were reportedly responsible for spreading misinformation about the pandemic and vaccinations (Abidin et al., 2021). However, at the same time, the Finnish government turned to SMIs to reach young people and offer them reliable information about the pandemic (Reinikainen et al., 2022). This shows how SMIs are often considered both a societal problem and a solution to societal issues.

2.2.3 The Participative and Perplexing Nature of Social Media

Interactivity is a central aspect that young people expect from digital and social media, as detailed in the previous section. Young people are found to be more likely to interact with news online compared with older people but are also more likely to get involved in 'signalling' activities, such as liking and sharing news stories online, instead of 'expressive' activities, such as commenting or posting opinions online (Sang et al., 2020). Depending on their objectives and target groups, media corporations use various online strategies (e.g., news content and platform choices, brand promotion, polls, contests, and asking for feedback) to elicit audience engagement (e.g., liking, sharing, commenting, posting emojis, and tagging friends) on social media platforms (Hase et al., 2023; Sormanen et al., 2022b). Social media has thus created more direct and quick forms of participation (e.g., direct feedback via liking or commenting on news) for online news media audiences compared with pre-digital media participation options (e.g., written snail mail or phone feedback to journalists). Hence, the technological affordances of social media platforms and the journalistic strategies in them, such as encouraging quick (visual) participation, can create heuristic shortcuts (see Sundar & Limperos, 2013) for young people because their needs and gratifications are in line with the low-deliberative and less time-consuming participation possibilities, or the digital and social platforms have made young people expect effortless participative content from online journalism.

Young people are also participating in other online journalistic content besides traditional news. Research from a Finnish teen magazine's Instagram audience showed that teen girls in particular want to engage with positively framed journalistic content on social media with quick likes and visual emojis and tags for peer support, sociability, and advocacy of societal issues (Sormanen et al., 2022b). The print magazine created an 'online audience community' on Instagram, which functioned as a venue for young people to receive information in mostly photo and video format, engage with the content, and contemplate and discuss various topics from everyday life to civic and political matters, and news among their peers. The online community also allowed young people to reflect their peers' interests and opinions through surveillance, hence giving them signs of what other teens considered worthy topics or behaviour. The online community was also beneficial to the magazine because it enabled gathering information on young audiences' needs and offering content on other social media platforms through links from the

Instagram page. The media strategy of the magazine was thus in line with all the MAIN Model's gratifications: mode, interactivity, agency, and navigability (see Sundar, 2008; Sundar & Limperos, 2013).

In many countries, public service media are reaching difficult-to-reach audiences, such as youngsters, through social media. For example, VRT, the public service media in Flanders (Belgium), produces news on social media platforms such as Instagram in native form (i.e., uniquely produced for the specific platform), referred to as 'social journalism', for young people in particular (Hendrickx, 2023). This social media information affordance has been noted as a popular venue for young people to stay in tune with various news topics and societal happenings in a form suitable for them and to engage with the content, also giving journalists indications of what news topics are of interest to the young. However, some traditional media choose not to engage with audiences on social media because of strategic choices (e.g., putting effort into their own websites and their traffic) or because they generally do not want to rely on big social media companies that dictate news flows and people's engagement possibilities with news strategically or through algorithms (e.g., Newman et al., 2023).

SMIs, on the other hand, rely strongly on information sharing and engagement with their audiences. While the functions and significance of SMIs and journalism are generally considered distinct from each other, the practices and roles of SMIs and journalists are becoming increasingly blurred in the new digital environment. Journalists often strike a balance between the personal and the professional and between the editorial and the commercial (Mellado & Hermida, 2022). Many journalists have turned to platforms such as TikTok and have adopted content logics more typical of SMIs (Negreira-Rey et al., 2022). Owing to the precarious nature of media work these days, many freelance journalists are also working as influencers and engage in influencer marketing activities (Frig & Penttilä, 2023) – a development that many consider worrisome due to the desire to ensure the integrity and trustworthiness of journalism.

At the same time, boundaries are also being stretched in the other direction, with SMIs experimenting with content types that can be considered journalism (Maares & Hanusch, 2020), participating in societal and political discussions (Suuronen et al., 2022), and calling out harmful practices by commercial brands (Frig & Penttilä, 2023), thus stepping into the territories of legacy media. For online audiences, this type of blurring of boundaries may be perplexing. Ehrlén et al. (2023) observed that young people often experience content, platform, and data confusion while online. Uncertainty about who is behind certain content and why, may affect trust in news and information providers and may even have profound consequences for democracy (Ehrlén et al., 2023).

2.3 DISCUSSION

This chapter shows how online news use and social media use (e.g., following SMIs) vary among young people but are profoundly guided by very similar motivations and gratifications that are incorporated into the processes of social and digital platforms. The research also shows that young people react positively to and engage with journalistic content online that abides by the U&G theory and the MAIN Model's features, such as visual content and interactivity. Some traditional news media have adapted to the expectations and desires of young people by providing them with short social media content on their favourite platforms, therefore also adopting online practices typical of SMIs.

2.3.1 Practical Implications

Digital environments can make determining the credibility of information more difficult than before. Young people need education in news media literacy skills, which refers to the knowledge and motivations required to identify and engage with journalism. Young news media-literate individuals understand how the news and media industry operates and its effects on people. They have the ability to critically assess information and news, especially in the current digital age, while still having the motivation to consume news (e.g., Ku et al., 2019). People around the world are expressing great concerns about their ability to recognise what constitutes credible news and information versus mis- and disinformation on social and digital media (Newman et al., 2023). Young people encounter information from many actors on social media, including legacy media and social media influencers (SMIs), and they particularly need help in learning to evaluate the credibility of the senders and the information they provide. Additionally, young people also need to understand how algorithms can guide their social media news consumption by suggesting news on feeds based on their past behaviour, and the negative consequences of this, such as information bubbles that distort and limit one's knowledge.

SMIs are tempting sources of news and information for young people because they can offer important reference points and companionship-like relationships. The number of their followers often works as cues, suggesting their credibility as information sources, which can lead to involuntary consumption of misinformation (e.g., Abidin et al., 2021; Sundar, 2008). Nonetheless, SMIs can provide young people with information, such as on civic and political issues, in a way that matches their needs and expectations. In addition, SMIs can provide access to the hard-to-reach youth segment, such as for media organisations.

The news industry and scholars have attempted to address the problem of increasing news avoidance among young people. Suggestions include making news more accessible for hard-to-reach groups, broadening the news agenda, commissioning more inspiring or positive news, or embracing more constructive journalism, solutions journalism, conciliatory journalism, or slow journalism to provide people with a greater sense of hope or personal agency regarding negative world phenomena. Other suggestions are reducing the cognitive costs of navigating the news environment, delivering customised news to prevent news overload, and increasing people's trust in journalism (Newman et al., 2023; Villi et al., 2021).

2.3.2 Avenues for Future Research

There are still potential risks and a need for future research on how this news media accommodation, through short, effortless online news, maintains journalism's values and influences young people's reading skills, news media literacy skills, and knowledge of societal and political issues. Moreover, if young people follow news only through accidental social media encounters or recommendations by SMIs, they may obtain a fragmented and limited picture of societal phenomena and essential news provided by traditional media. On the positive side, news media content and visibility on social media, even in short visual formats, can enable young people to receive more accurate information and feel that they are also a part of the journalistic information sphere, in addition to the social media information sphere. Journalistic development and research are needed to find a balance between online journalistic content that follows the rules and values of journalism, appeals to young people's wants, and coincides with young people's development and learning needs to help them grow up as informed, literate citizens.

Further challenges and areas in need of research include the development of artificial intelligence (AI), which makes it more difficult to check the accuracy of online content. For example, the MAIN Model's modality-based gratifications (e.g., 'realism heuristic') may be changed due to AI because picture and video content circulated online can no longer be judged straightforwardly as trustworthy, regardless of the sources. Overall, journalistic media face a great challenge in maintaining trust among young audiences as sources of reliable content in today's and the future's information-full and continuously developing online media landscape.

AUTHORS' NOTE

This work was supported by the Academy of Finland (#353964).

REFERENCES

Abidin, C. (2016). 'Aren't these just young, rich women doing vain things online?': Influencer selfies as subversive frivolity. *Social Media + Society, 2*(2), 1–17. https://doi.org/10.1177/2056305116641342.

Abidin, C., Lee, J., Barbetta, T., & Miao, W. S. (2021). Influencers and COVID-19: Reviewing key issues in press coverage across Australia, China, Japan, and South Korea. *Media International Australia, 178*(1), 114–135. https://doi.org/10.1177/1329878X20959838.

Boczkowski, P. J., Mitchelstein, E., & Matassi, M. (2018). News comes across when I'm in a moment of leisure: Understanding the practices of incidental news consumption on social media. *New Media & Society, 20*(10), 3523–3539. https://doi.org/10.1177/1461444817750396.

Croes, E., & Bartels, J. (2021). Young adults' motivations for following social influencers and their relationship to identification and buying behavior. *Computers in Human Behavior, 124*, 106910. https://doi.org/10.1016/j.chb.2021.106910.

Edgerly, S., Vraga, E. K., Bode, L., Thorson, K., & Thorson, E. (2018). New media, new relationship to participation? A closer look at youth news repertoires and political participation. *Journalism & Mass Communication Quarterly, 95*(1), 192–212. https://doi.org/10.1177/1077699017706928.

Ehrlén, V., Talvitie-Lamberg, K., Salonen, M., Koivula, M., Villi, M., & Uskali, T. (2023). Confusing content, platforms, and data: Young adults and trust in news media. *Media and Communication, 11*(4), 320–331. https://doi.org/10.17645/mac.v11i4.7104.

Enke, N., & Borchers, N. S. (2019). Social media influencers in strategic communication: A conceptual framework for strategic social media influencer communication. *International Journal of Strategic Communication, 13*(4), 261–277. https://doi.org/10.1080/1553118X.2019.1620234.

Frig, M-M., & Penttilä, V. (2023). Toimittaja-vaikuttajat organisaatioiden vastuullisuusviestinnän sidosryhmänä [Journo-influencers as stakeholders in organizations' sustainability communication] In H. Reinikainen & S-M. Laaksonen (Eds), *ProComma Academic 2023: Vaikuttava viestintä* [*ProComma Academic 2023: Influential communication*] (pp. 113–130). ProCom – Viestinnän ammattilaiset ry. https://doi.org/10.31885/9789526523910.

Galan, L., Osserman, J., Parker, T., & Taylor, M. (2018). *How young people consume news and the implications for mainstream media.* A report by Flamingo commissioned by the Reuters Institute for the Study of Journalism, Oxford University. https://reutersinstitute.politics.ox.ac.uk/sites/default/files/2021-02/FlamingoxREUTERS-Report-Full-KG-V28.pdf.

Geers, S. (2020). News consumption across media platforms and content: A typology of young news users. *Public Opinion Quarterly, 84*, 332–354. https://doi.org/10.1093/poq/nfaa010.

Gil de Zúñiga, H., Weeks, B., & Ardèvol-Abreu, A. (2017). Effects of the news-finds-me perception in communication: Social media use implications for news seeking and learning about politics. *Journal of Computer-Mediated Communication, 22*, 105–123. https://doi.org/10.1111/jcc4.12185.

Hase, V., Boczek, K., & Scharkow, M. (2023). Adapting to affordances and audiences? A cross-platform, multi-modal analysis of the platformization of news on Facebook,

Instagram, TikTok, and Twitter. *Digital Journalism*, *11*(8), 1499–1520. https://doi.org/10.1080/21670811.2022.2128389.

Hasell, A., & Chinn, S. (2023). The political influence of lifestyle influencers? Examining the relationship between aspirational social media use and anti-expert attitudes and beliefs. *Social Media + Society*, *9*(4). https://doi.org/10.1177/20563051231211945.

Hendrickx, J. (2023). The rise of social journalism: An explorative case study of a youth-oriented Instagram news account. *Journalism Practice*, *17*(8), 1810–1825. https://doi.org/10.1080/17512786.2021.2012500.

Horton, D., & Wohl, R. R. (1956). Mass communication and para-social interaction: Observation on intimacy at a distance. *Psychiatry*, *19*(3), 215–229. https://doi.org/10.1080/00332747.1956.11023049.

Huang, E. (2009). The causes of youths' low news consumption and strategies for making youths happy news consumers. *Convergence*, *15*(1), 105–122. https://doi.org/10.1177/1354856508097021.

Katz, E., Blumler, J. G., & Gurevitch, M. (1973–1974). Uses and gratifications research. *Public Opinion Quarterly*, *37*(4), 509–523.

Ku, K. Y. L., Kong, Q., Song, Y., Deng, L., Kang, Y., & Hu, A. (2019). What predicts adolescents' critical thinking about real-life news? The roles of social media news consumption and news media literacy. *Thinking Skills and Creativity*, *33*, 100570. https://doi.org/10.1016/j.tsc.2019.05.004.

Lee, H., & Yang, J. (2014). Political knowledge gaps among news consumers with different news media repertoires across multiple platforms. *International Journal of Communication*, *8*, 597–617. https://ijoc.org/index.php/ijoc/article/view/2455.

Lunt, P., & Vicente-Mariño, M. (2015). From spectators to users: Exploring the ongoing transformation of media audiences. *Trípodos*, *36*, 9–11. http://www.tripodos.com/index.php/Facultat_Comunicacio_Blanquerna/article/view/239.

Maares, P., & Hanusch, F. (2020). Exploring the boundaries of journalism: Instagram micro-bloggers in the twilight zone of lifestyle journalism. *Journalism*, *21*(2), 262–278. https://doi.org/10.1177/1464884918801400.

Martin, J. D., & Sharma, K. (2022). Getting news from social media influencers and from digital legacy news outlets and print legacy news outlets in seven countries: The 'more-and-more' phenomenon and the new opinion leadership. *Newspaper Research Journal*, *43*(3), 276–299. https://doi.org/10.1177/07395329221105507.

Mellado, C., & Hermida, A. (2022). A conceptual framework for journalistic identity on social media: How the personal and professional contribute to power and profit. *Digital Journalism*, *10*(2), 284–299. https://doi.org/10.1080/21670811.2021.1907203.

Negreira-Rey, M.-C., Vazquez-Herrero, J., & Lopez-García, X. (2022). Blurring boundaries between journalists and tiktokers: Journalistic role performance on TikTok. *Media and Communication*, *10*(1), 146–156. https://doi.org/10.17645/mac.v10i1.4699.

Newman, N., Fletcher, R., Eddy, K., Robertson, C. T., & Kleis Nielsen, R. (2023). *Reuters Institute Digital News Report 2023*. Reuters University, University of Oxford. https://reutersinstitute.politics.ox.ac.uk/digital-news-report/2023.

Nielsen, R. K. (2016). The many crises of western journalism: A comparative analysis of economic crises, professional crises, and crises of confidence. In J. C. Alexander, B. E. Butler, & M. Luengo (Eds), *The Crisis of Journalism Reconsidered* (pp. 77–97). Cambridge University Press.

Pöyry, E., & Reinikainen, H. (2024). 'Remember to vote!' – How do people respond to social media influencers promoting political participation? In J. Arnesson & H.

Reinikainen (Eds), *Influencer Politics* (pp. 35–50). De Gruyter. https://doi.org/10.1515/9783111036106-004.
Reinikainen, H., Laaksonen, S.-M., Pöyry, E., & Luoma-aho, V. (2022). Conquering the liminal space: Strategic social media influencer communication in the Finnish public sector during the COVID-19 pandemic. In O. Niininen (Ed.), *Social Media for Progressive Public Relations* (pp. 161–175). Routledge. https://doi.org/10.4324/9781003177791-15.
Reinikainen, H., Munnukka, J., Maity, D., & Luoma-aho, V. (2020). 'You really are a great big sister' – Parasocial relationships, credibility, and the moderating role of audience comments in influencer marketing. *Journal of Marketing Management, 36*(3–4), 279–298. https://doi.org/10.1080/0267257X.2019.1708781.
Sang, Y., Lee, J. Y., Park, S., Fisher, C., & Fuller, G. (2020). Signalling and expressive interaction: Online news users' different modes of interaction on digital platforms. *Digital Journalism, 8*(4), 467–485. https://doi.org/10.1080/21670811.2020.1743194.
Schanke Sundet, V., & Lüders, M. (2023) 'Young people are on YouTube': Industry notions on streaming and youth as a new media generation. *Journal of Media Business Studies, 20*(3), 223–240. https://doi.org/10.1080/16522354.2022.2125262.
Schrøder, K. C., Blach-Ørsten, M., & Kæmsgaard Eberholst, M. (2020). Is there a Nordic news media system? A descriptive comparative analysis of Nordic news audiences. *Nordic Journal of Media Studies, 2*, 23–35. https://doi.org/10.2478/njms-2020-0003.
Skovsgaard, M., & Andersen, K. (2020). Conceptualizing news avoidance: Toward a shared understanding of different causes and potential solutions. *Journalism Studies, 21*(4), 459–476. https://doi.org/10.1080/1461670X.2019.1686410.
Sormanen, N., Rantala, E., Lonkila, M., & Wilska, T-A. (2022a). Moderate digital traditionalists, minimalist social media stumblers and frequent omnivores: News consumption repertoires among Finnish adolescents. *Nordicom Review, 43*(2), 234–253. https://doi.org/10.2478/nor-2022-0014.
Sormanen, N., Reinikainen, H., & Wilska, T-A. (2022b). Strategies of eliciting young people's affective and quick participation in a youth magazine's Instagram community. *Journalism Practice, 17*(8), 1790–1809. https://doi.org/10.1080/17512786.2021.2016070.
Sundar, S. S. (2008). The MAIN model: A heuristic approach to understanding technology effects on credibility. In M. J. Metzger & A. J. Flanagin (Eds), *Digital Media, Youth, and Credibility* (pp. 72–100). The MIT Press. https://doi.org/10.1162/dmal.9780262562324.073.
Sundar, S. S., & Limperos, A. M. (2013). Uses and grats 2.0: New gratifications for new media. *Journal of Broadcasting & Electronic Media, 57*(4), 504–525. https://doi.org/10.1080/08838151.2013.845827.
Suuronen, A., Reinikainen, H., Borchers, N. S., & Strandberg, K. (2022). When social media influencers go political: An exploratory analysis on the emergence of political topics among Finnish influencers. *Javnost – The Public, 29*(3), 301–317. https://doi.org/10.1080/13183222.2021.1983367.
Villi, M., Aharoni, T., Tenenboim-Weinblatt, K., Boczkowski, P. J., Hayashi, K., Mitchelstein, E., Tanaka, A., & Kligler-Vilenchik, N. (2021). Taking a break from news: A five-nation study of news avoidance in the digital era. *Digital Journalism, 10*(1), 148–164. https://doi.org/10.1080/21670811.2021.1904266.

Vozab, D. (2019). Generational patterns of digital news consumption: From traditionalists to millennial minimalists. *Medijske studije*, *10*(20), 107–126. https://doi.org/10.20901/ms.10.20.6.

Yoo, C. H. (2011). Modeling audience interactivity as the gratification-seeking process in online newspapers. *Communication Theory*, *21*, 67–89. https://doi.org/10.1111/j.1468–2885.2010.01376.x.

3. Growing up together with artificial intelligence: distribution of agency in commercial digital environments

Jussi Nyrhinen and Eero Rantala

3.1 INTRODUCTION

The purpose of this chapter is to shed light on how artificial intelligence (AI) has created a new way to practise agency but can also reinforce structures. Notably, young people are leading the change in consumption in commercial digital environments, such as social media sites, online stores, streaming and downloading services, and search engines, because they are frequent users of these environments (Nyrhinen et al., 2024). Thus, young people are active agents in these environments (Rissanen & Luoma-aho, 2016; Reinikainen et al., 2020). This means that they can act independently and make their own free choices within such environments and/or that digital environments may facilitate their agency (Willett, 2008). However, agency is practised within structures such as consumer culture and markets (Appadurai, 1996; Kenway & Bullen, 2001). Consequently, these structures also constrain agency (Willett, 2008).

Like any previous generation, people who are currently in their late teens or young adulthood (twenties or thirties) have grown up with technologies regarded as novel by prior generations. For people born after 1984 (or after 1990, see Kärki, 2024), this concerns particularly advanced digital technologies, which is why these people are categorised as 'diginatives' (Kirschner & De Bruyckere, 2017). However, it is justified to say that unlike any other prior technology, the current AI technologies not only shape their users, but their functioning is also continuously shaped by their users. The reason for this is that the function of AI relies on data that consumers constantly engender by leaving tracks of their behaviours in digital environments (Liu & Wei, 2021; Nyrhinen et al., 2024; Varnali, 2021).

In many commercial digital environments, AI receives data from people who are interacting with it (Boerman et al., 2017; Chen & Stallaert, 2014;

Nyrhinen et al., 2024; Varnali, 2021). Subsequently, consumer data are utilised in planning the production and targeting the offerings, which in turn steer consumer behaviour (Dodoo & Wu, 2019; Kalyanaraman & Sundar, 2006; Nyrhinen et al., 2024). Therefore, online agency can be regarded as distributed between the technology and people. However, in this cycle lies a risk of stagnation or 'becoming embalmed' (see Paavola & Cuthbertson, 2021) caused by a feedback loop. Such a loop is created when consumers follow the AI recommendations, and their behaviour generates data that are used to teach AI further. From the consumer's perspective, the feedback loop can be seen as stagnation. This is because the loop causes AI recommendations to excessively favour alternatives that are already popular or have been previously chosen by the same consumer while neglecting alternatives that could better match the consumer's actual needs and preferences. In this chapter, we discuss issues related to stagnation from the perspective of young people's agency.

A common everyday application of AI is online behavioural targeting in marketing, in which businesses use data gathered from consumers' web browsing and search activities, as well as other publicly available online information, to show them more relevant ads or product recommendations, including content recommendations on various streaming and downloading platforms (Boerman et al., 2017; Chen & Stallaert, 2014; Nyrhinen et al., 2024; Varnali, 2021). Furthermore, to improve user engagement and send consumers relevant and customised information, websites and applications track users' online actions to deduce their preferences (Liu & Wei, 2021).

It is important to examine the impact of AI on young people's agency because it has been increasingly used in several fields during the current and past decades thanks to the ever-growing computational capacity (Russell & Norvig, 2020). Although the basic principle behind AI has existed for several decades, only in the past few years have various AI applications been widely adopted by consumers and marketers alike. In particular, the use of AI-driven targeted marketing has been steadily increasing, and young people are among the prime targets of corporations and institutions (Aragoncillo & Orus, 2018; Ruckenstein & Granroth, 2020). As a result, young people are heavily exposed to internet marketing (Dhanesh & Duthler, 2019) and are vulnerable to peer pressure and social media norms (Savolainen et al., 2021). Simultaneously, young people develop habits, identities, and roles as active social agents with consumption (Willett, 2008; Deutsch & Theodorou, 2010; Wilska, 2017), which involves choosing products to define themselves as members of specific social groups and distinguishing themselves from other groups (e.g., Hebdige, 1979; Lury, 1996). In addition, the idea of diginatives being tech- and/or information-savvy in contrast to prior generations is a myth that has long been debunked (see, for example, Kirschner & De Bruyckere, 2017; Kärki, 2024). Rather, there is evidence that diginatives are often unable to use new technologies in the most efficient or effective manner because they tend to fail to recognise the enhanced functionalities of the applications (Bullen et al., 2008; Kirschner & De Bruyckere, 2017). This is because they have only experienced a digitally connected world and often passively utilise the technologies as they are given to them (Kirschner & De Bruyckere, 2017).

Although there has been research on AI from the technical and managerial perspectives (e.g., Mansoury et al., 2020), there is limited knowledge on it from the social perspective. Prior research has examined young people's online agency in relation to commercial structures by examining how they resist, redefine, and recontextualise, and how they are being positioned by cultural products and discourses online (Willett, 2008). Paavola and Cuthbertson (2021), for instance, have examined how AI shifts agency from the perspective of organisational structures. In addition, the distribution of young agency between people in social media has also been examined (e.g., Enfield & Kockelman, 2017; see also Glăveanu, 2015, for 'co-agency'). However, to the best of our knowledge, research on the distribution of agency between AI and young people is scarce. Therefore, the aim of this chapter is to describe how AI both supports and hinders young people's agency. To achieve this aim, we conducted a technology analysis and theoretical conceptualisation, relying on the existing literature. We contribute to the theory of agency in several ways. First, we explain how the use of consumer data in behavioural targeting distributes agency by automatically engaging consumers in corporate decision-making. Second, we discuss issues related to stagnation engendered by biases in data from the perspective of young people's agency. Third, we explain how adjusting AI may intentionally or unintentionally misguide young people online and how AI reduces young people's agency.

3.2 AI AND ONLINE BEHAVIOURAL TARGETING

Behavioural targeting, online behavioural advertising, or online targeted advertising is the technique of showing customers tailored adverts based on tracking their online activity (Boerman et al., 2017; Chen & Stallaert, 2014; Liu & Wei, 2021; Nyrhinen et al., 2024; Varnali, 2021). Targeted advertising and other algorithm-based recommendations are mostly determined by the user's past online behaviour and preferences (Chen & Stallaert, 2014; Liu & Wei, 2021; Nyrhinen et al., 2024) as well as those of their online peers on social media (Bakshy et al., 2012; Nyrhinen et al., 2024). This type of data gathered by various online platforms is regarded as a manifestation of big data. Typical for big data is that such data update constantly, have a vast number of statistical units and/or units of observation, and are often unstructured, which means that big data can contain various kinds of data, such as texts, numbers, and images (Russell & Norvig, 2020, 1–30; Mohri et al., 2018, 1–8; Goodfellow et al., 2016, 1–28).

AI is employed in behavioural targeting to track users and draw conclusions about their attributes, inclinations, search histories, media consumption information, purchases, and online browsing habits (Liu & Wei, 2021; Nyrhinen et al., 2024; Varnali, 2021). It is necessary to effectively analyse these vast, constantly updating, and unstructured data because conventional methods require more structured data. Reciprocally, teaching AI requires a massive amount of data. The type of AI used to analyse big data is usually deep learning, in which AI learns from data (i.e., multiple examples) to perform different tasks (Russell & Norvig, 2020, 651–714). Therefore, small datasets rarely provide enough examples to adequately teach AI. The goal of deep learning is to teach a programme to perform a given task and possible future tasks with different datasets (Russell & Norvig, 2020, 651–714; Goodfellow et al., 2016, 110–131, 228–271). A deep learning neural network consists of input and output layers with multiple hidden layers between the two, which help in performing tasks more accurately than more superficial neural networks can (Russell & Norvig, 2020, 750–788; Goodfellow et al., 2016, 1–28, 166–227). Deep learning models can take data into account in real time, and models are continuously adjusted with new data to perform their tasks more efficiently and accurately (Steck et al., 2021; Russell & Norvig, 2020, 27–30; Goodfellow et al., 2016, 478–482). For these reasons, AI and big data are intertwined. By observing users' online actions, online behavioural targeting aims to increase user engagement and display more relevant advertisements (Boerman et al., 2017; Chen & Stallaert, 2014; Liu & Wei, 2021; Nyrhinen et al., 2024; Varnali, 2021).

Online advertisements are tailored to individual users with deep learning by considering their online personas, online surfing patterns, and interests (Nyrhinen et al., 2024). This type of advertising is based on deep learning models known as *recommendation systems* (Hussien et al., 2021). For instance, search engine advertisements or recommendations by streaming services are based on such systems. The task of recommender systems is to recommend adequate content for consumers (e.g., entertainment, product advertisements). As input data, recommender systems take consumer behaviour data, which can be texts, images, audio, and videos (Steck et al., 2021; Russell & Norvig, 2020, 27–30; Goodfellow et al., 2016, 478–482). Recommendations can be based on similarities between consumers (collaborative filtering), individual users' activities (content-based filtering), or often a combination of these two (Goyani & Chaurasiya, 2020). This allows recommendations to be matched to user demands and interests (Dodoo & Wu, 2019; Goyani & Chaurasiya, 2020). Consequently, consumers may view tailored ads as more relevant to them because these align with some aspects of their personality (Dodoo & Wu, 2019; Kreuter & Wray, 2003; Petty et al., 2002), which can be regarded as a form of self-referencing (Nyrhinen et al., 2024). Earlier studies have shown that perceived relevance is the factor that explains the effectiveness of personalisation (e.g., Dodoo & Wu, 2019; Kalyanaraman & Sundar, 2006).

3.3 DISTRIBUTION OF AGENCY AND STAGNATION

3.3.1 Distributed Agency between Humans and AI

There is tension between the ideas of young people's online actions; on the one hand, young people are regarded as independent actors, while on the other hand, they are seen as acted upon by societal forces online (Willett, 2008). A unified view by Willett (2008) portrays young people as 'bricoleurs' surrounded by market structures who appropriate and reshape consumer culture as they define and act out their identities or alternatives, ignoring these. That consumers are passive objects of online marketing is a false assumption, but prior literature also avoids overemphasising freedom of choice online (Willett, 2008). Much of the interaction between young people and companies takes place in digital environments that are easy to manipulate (Reinikainen et al., 2020). Digital spaces are technologically created, often by commercial actors. Therefore, it can be said that structures are encoded in these environments. This means that, besides other societal structures and actors, merchandising shapes consumers' freedom of choice and how they act upon it. Merchandising in such environments may appear so subtly that consumers regard themselves as actors when they are, at most, 'choosers' acting within marketers' terms (see Appadurai, 1996). However, online marketing decision-making is integrated into AI due to behavioural targeting. Thus, power and accountability are partially shifted to consumers, which blurs who oversees a system steered by AI (Paavola & Cuthbertson, 2021). More precisely, with the use of AI, consumers can be regarded as no longer solely customers or end users but as co-decision-makers on production and supply matters, although their power is opaque (Paavola & Cuthbertson, 2021). Hence, e-consumer behaviour has become a major component in online marketing decision-making (Paavola & Cuthbertson, 2021).

3.4 BIASED DATA AND FEEDBACK LOOPS

AI's reliance on data makes it prone to 'data bias' (Olteanu et al., 2019) or 'training data bias' (Danks & London, 2017). This means that if AI is taught with biased data, its outputs will also be biased (Danks & London, 2017; Olteanu et al., 2019). Even with careful adjustment of AI and datasets, it may not be possible to eliminate bias entirely. AI learns from historical data, which encodes historical biases (Roselli et al., 2019). This issue is exacerbated when

AI's output impacts its input, creating a vicious cycle of bias that deteriorates the application by affecting its predictive performance indefinitely (Ensign et al., 2018; Roselli et al., 2019). If consumers are very prone to following recommendations, this can lead in time and with an amplifying loop to a situation where only a small portion of different products or media content is recommended, consumed, and rated (Mansoury et al., 2020). Mansoury et al. (2020) referred to this phenomenon as a feedback loop in which consumer data are continuously generated to adjust the system in real time (see also Abdollahpouri et al., 2020; Chen et al., 2023).

3.5 POPULARITY BIAS

One of the most central consequences of stagnation caused by a feedback loop is that it amplifies popularity bias. AI is biased towards cases that more commonly occur in the training data and is thus less likely to take rare cases into account (Mac Namee et al., 2002). This means that a recommendation system suggests only the most popular items to consumers (Mansoury et al., 2020; Abdollahpouri et al., 2020). The origin of the popularity bias is in the initial training data, which consists of a few products that are popularly purchased or reviewed – that is, the distributions of purchases and reviews are 'organically' skewed towards the most popular. This issue is amplified when consumers follow recommendations by AI and thereby produce more data that are increasingly skewed towards the most popular cases (Mansoury et al., 2020; Abdollahpouri et al., 2020).

A recommender system such as the one described above can be regarded as unfair because its recommendations do not fairly represent the tastes of one group of consumers, while the other group of consumers receives recommendations that match their actual preferences (Abdollahpouri et al., 2020). Adequately working recommendation systems should take into account different consumers and should be able to recommend niche items for consumers whose tastes differ from those of the majority (Mansoury et al., 2020; Abdollahpouri et al., 2020). A recommended list of items may be considered good if it is relevant to the user, is diverse, and helps the user discover products they would not have not been able to discover in the absence of the recommender system (Abdollahpouri et al., 2020).

Popularity bias is among the issues that deteriorate agency online. This is because it mitigates an individual's impact on the products and content suggested to them by recommendation machines. Instead of offering relevant, diverse, or novel alternatives, popularity bias suggests products (including media content) that are already favoured by many and therefore reoccur in the training data and in suggestions (Abdollahpouri et al., 2020). In this sense, rather than enhancing agency, technology shifts power to structures such as merchandising and the dominant consumer culture (see Willett, 2008).

To conclude, feedback loops and the subsequent popularity bias lead to stagnation, which hinders consumers' agency. While the ideal intended purpose of recommendation systems is to help consumers discover novel and diverse selections of relevant products and content, loops and biases lead recommendation systems to suggest a limited set of items that are already popular or have been previously chosen by the same consumer. It can be said that this stagnation limits consumers' ability to practise agency online and shifts power back to structures. This is because it reinforces the popularity of products and contents that have already gained much attention or a large market share.

3.6 DEVELOPER BIAS

3.6.1 Challenges in Mitigating Data Bias

Developers may try to mitigate data bias by adjusting a recommendation system so that it can provide more adequate recommendations to its users. This means that the results of a recommendation system would be based on consumers' and their peers' true behaviours and would be interfered with as little as possible. In such a scenario, consumers' behaviours could affect the recommendations they receive. However, adjusting a recommendation system can engender another type of bias: *developer bias*. This type of bias occurs when a programme developer modifies a recommendation system to alter its outcomes. Despite the intention to fix the system so that it would produce more relevant results, these results may not be based on the consumers' behaviours and AI's function but on the developer's choices and actions. Therefore, the developer's involvement can reduce the consumers' agency because it shifts power from the consumers to the marketers. Developer bias is always context-specific and related to the contextually defined aim or task of the recommendation system and its users (both consumers and developers). In addition, adjusting the models does not always lead to bias (Mansoury et al., 2020; Abdollahpouri et al., 2020; Chen et al., 2023). For example, recommendation systems that have managed to mitigate, to some extent, bias amplification originating from feedback mechanisms or popularity bias have been adjusted (Mansoury et al., 2020; Abdollahpouri et al., 2020; Chen et al., 2023). These kinds of adjustments cannot be defined as biased in a given context.

3.7 ALTERING RECOMMENDATION SYSTEMS

In addition to failed attempts to improve recommendation systems, developers may intentionally create bias. A common reason for this is revenue models for several commercial digital environments, such as commercial platforms, search engines, downloading and streaming platforms, and social networking sites that consumers use to search for products and content. An example of such a revenue model is sponsored search advertising or keyword advertising. In such a revenue model, advertisers are allowed to bid for their query-specific advertisements to appear alongside organic search results (Yuan et al., 2017). More importantly, advertisers may also affect the visibility of their products or content within seemingly organic search results (Bhattacharya et al., 2024). Alongside the popularity bias, this leads to the issue that recommendation systems do not suggest products or content that are relevant to a consumer based on their choices and individual preferences. Rather, those products and content that have gained past exposure, due to their first-exposed advantage or paid rankings, gain even more future exposure (Patro et al., 2022). Therefore, supervisor bias (either intentional or unintentional) can be seen as a factor that hinders individual agency by making consumers' choices more structured by commercial mechanisms such as marketing (see Willett, 2008).

3.8 DISCUSSION

This chapter describes how AI may both increase and decrease the agency of consumers in commercial digital environments. Prior research (i.e., Paavola & Cuthbertson, 2021) has acknowledged that unintentional and uncontrollable consumer behaviours linked with marketing decision-making may cause issues from a corporate perspective. This chapter, however, is among the first to discuss in depth the opportunities and threats of AI recommendations from the perspective of consumers' agency. Because advertising's behavioural targeting is based on tracking consumers' actions online, consumers practise agency in a new, often unintentional, automated manner. However, because AI is dependent on given data, biases in data and feedback loops may distort recommendations so that they do not reflect a consumer's actual behaviour. In addition, recommendations based on limited data collected from one consumer or a small group may be technically accurate but useless to a consumer because they provide minimal novel information. Such recommendations may also lead to a filter bubble (Aridor et al., 2020), meaning that they only reinforce consumers' existing attitudes and practices. In sum, besides introducing how the use of AI has engendered a new way to practise agency, this chapter highlights issues that may even reinforce structures that frame consumer choice. More importantly, consumers unknowingly cede more of their agency if they perceive AI recommendations as genuine representations of their personal tastes and preferences.

The aforementioned issues particularly concern young people for several reasons. First, young people spend more time in online environments and are thus are more affected by behavioural targeting and content recommendations. Second, young people socialise as consumers and develop their agency and personalities. Cultural content products such as books and audiovisual content may significantly affect their values, opinions, and common knowledge. By following AI's recommendations without much questioning, young people may shift their agency to marketers and other institutions that aim to influence their attitudes and behaviours. Owing to the self-referencing nature of online recommendations, such information may appear relevant and personal, although it may be affected by various biases and particularly marketers' intentions. Issues such as the filter bubble may not only limit young consumers' choices but may also exacerbate the polarisation of tastes, attitudes, and behaviours between different groups of people.

3.9 PRACTICAL IMPLICATIONS

Based on our conceptualisation, practical implications can be drawn from the consumer, corporate, and societal perspectives. Owing to the biased suggestions made by recommendation systems, consumers may not be aware of the best alternatives (e.g., products that are of the best quality, are the most affordable, and are tailored for their purposes). This issue may become more severe if consumers lack digital media literacy and assume that suggestions made by recommendation systems are genuine representations of their tastes and preferences. Therefore, young people should be taught digital media literacy early on to help consumers acknowledge how various recommendation systems are intentionally or unintentionally biased. In addition, educators and various institutions that advocate consumers' and young people's rights should help tackle issues related to biased AI recommendations. One way to do this is to try to affect self-regulation in the fields of information technology and commerce, regardless of how challenging it may be.

Second, from a commercial perspective, automated targeted marketing may often be more profitable than mass marketing. More importantly, it can enable sellers to cut down costs and address environmental issues related to waste and unnecessary logistics. For sellers on various online commercial platforms, both data bias and developer bias create inequality. Novel and niche products and content, in particular, may be overshadowed by those that have a first-exposure advantage, are favoured by the masses, and/or are paid for so that they would reach the top of the rankings. This concerns not only consumers who would benefit from the availability of such niche items but also those citizens who work as micro-entrepreneurs on online platforms. These issues should be acknowledged by lobbying organisations and competition regulators, who have the power to affect the fairness of trade.

Third, because consumption has a drastic impact on how young people shape their identities, agency, and habits, bias in AI recommendation systems is also a societal concern. Many products consumed online are content, such as books, videos, and recordings, including news and documentaries. Such cultural products have a significant impact when consumers, especially the young, form their values and opinions. Moreover, several commercial digital environments are used for non-commercial purposes, such as information searches and social networking. Nevertheless, the bias in AI also affects searches that are not directly related to consumption and it similarly constrains what information consumers may obtain. Alongside commercial actors, various foreign and domestic political actors, such as political parties, lobbying organisations, and government agencies, are practising online behavioural targeting in their communications. They can also distort recommendation systems to serve their intentions or be treated unequally due to biases in recommendation systems.

3.10 AVENUES FOR FUTURE RESEARCH

This conceptualisation relied on secondary data to illustrate how AI-based recommendation systems affect young consumers' agency online. On one hand, our review helps set a future research agenda to further investigate how AI provides consumers with a new way to practise agency. On the other hand, we illustrated the concept of stagnation in recommendation systems and discussed how it is related to biases in AI, which should be acknowledged in future studies examining structures in digital environments. We focused on young people in a commercial context, but future research could examine these issues in other areas, such as research, education, and finance. For such studies, our conceptualisation provides a framework for dissecting the mechanisms by which AI affects people's agency. Due to the relative novelty of the phenomena studied, there is limited evidence of how severely biases and stagnation of online recommendations affect consumers' behaviour and attitudes. Therefore, alongside business cases and technology analysis, more consumer research, such as surveys, interviews, and observation studies, could be conducted.

AUTHORS' NOTE

This work was supported by the Strategic Research Council (SRC) at the Academy of Finland (#352544).

REFERENCES

Abdollahpouri, H., Mansoury, M., Burke, R., & Mobasher, B. (2020, September). The connection between popularity bias, calibration, and fairness in recommendation. In *Proceedings of the 14th ACM conference on recommender systems* (pp. 726–731).

Appadurai, A. (1996). *Modernity at large: Cultural dimensions of globalization* (Vol. 1). Minnesota: University of Minnesota Press.

Aragoncillo, L., & Orus, C. (2018). Impulse buying behaviour: An online–offline comparative and the impact of social media. *Spanish Journal of Marketing – ESIC, 22*(1), 42–62.

Aridor, G., Goncalves, D., & Sikdar, S. (2020, September). Deconstructing the filter bubble: User decision-making and recommender systems. In *Proceedings of the 14th ACM conference on recommender systems* (pp. 82–91).

Bakshy, E., Eckles, D., Yan, R., & Rosenn, I. (2012, June). Social influence in social advertising: Evidence from field experiments. In *Proceedings of the 13th ACM conference on electronic commerce* (pp. 146–161).

Bhattacharya, A., Robinson, S., Mookherjee, S., & Blöte, H. (2024). Examining the impact of sponsored search results on choice: An anchoring perspective. *Journal of Interactive Marketing, 59*(2), 183–200.

Boerman, S. C., Kruikemeier, S., & Zuiderveen Borgesius, F. J. (2017). Online behavioral advertising: A literature review and research agenda. *Journal of Advertising, 46*(3), 363–376.

Bullen, M., Morgan, T., Belfer, K., & Qayyum, A. (2008). *The digital learner at BCIT and implications for an e-strategy*. EDEN Research Workshop.

Chen, J., Dong, H., Wang, X., Feng, F., Wang, M., & He, X. (2023). Bias and debias in recommender system: A survey and future directions. *ACM Transactions on Information Systems, 41*(3), 1–39.

Chen, J., & Stallaert, J. (2014). An economic analysis of online advertising using behavioral targeting. *MIS Quarterly, 38*(2), 429–450.

Danks, D., & London, A. J. (2017). Algorithmic bias in autonomous systems. In *Proceedings of the 26th International joint conference on artificial intelligence* (IJCAI Vol. 17, No. 2017, pp. 4691–4697).

Deutsch, N. L., & Theodorou, E. (2010). Aspiring, consuming, becoming: Youth identity in a culture of consumption. *Youth & Society, 42*(2), 229–254.

Dhanesh, G. S., & Duthler, G. (2019). Relationship management through social media influencers: Effects of followers' awareness of paid endorsement. *Public Relations Review, 45*(3), 101765.

Dodoo, N. A., & Wu, L. (2019). Exploring the anteceding impact of personalised social media advertising on online impulse buying tendency. *International Journal of Internet Marketing and Advertising, 13*(1), 73–95.

Enfield, N. J., & Kockelman, P. (eds) (2017) *Foundations of human interaction: Distributed agency*. Oxford: Oxford University Press.

Ensign, D., Friedler, S. A., Neville, S., Scheidegger, C., & Venkatasubramanian, S. (2018). Runaway feedback loops in predictive policing. In *Conference on fairness, accountability and transparency* (pp. 160–171). PMLR.

Glăveanu, V. P. (2015). From individual agency to co-agency. In *Constraints of agency: Explorations of theory in everyday life* (pp. 245–265). Cham: Springer International Publishing.

Goodfellow, I., Bengio, Y., & Courville, A. (2016). *Deep learning*. Cambridge, MA: The MIT Press.
Goyani, M., & Chaurasiya, N. (2020). A review of movie recommendation system: Limitations, survey and challenges. *ELCVIA: Electronic letters on computer vision and image analysis*, *19*(3), 18–37.
Hebdige, D. (1979). *Subculture: The meaning of style*. London: Methuen.
Hussien, F. T. A., Rahma, A. M. S., & Wahab, H. B. A. (2021). Recommendation systems for e-commerce systems: An overview. *Journal of Physics: Conference Series*, *1897*(1), 012024.
Kalyanaraman, S., & Sundar, S. S. (2006). The psychological appeal of personalized content in web portals: Does customization affect attitudes and behavior? *Journal of Communication*, *56*(1), 110–132.
Kärki, K. (2024). Digital distraction, attention regulation, and inequality. *Philosophy & Technology*, *37*(1), 1–21.
Kenway, J., & Bullen, E. (2001). *Consuming children: Education-entertainment-advertising*. Buckingham: Open University Press.
Kirschner, P. A., & De Bruyckere, P. (2017). The myths of the digital native and the multitasker. *Teaching and Teacher Education*, *67*, 135–142.
Kreuter, M. W., & Wray, R. J. (2003). Tailored and targeted health communication: Strategies for enhancing information relevance. *American Journal of Health Behavior*, *27*(1), S227–S232.
Liu, B., & Wei, L. (2021). Machine gaze in online behavioral targeting: The effects of algorithmic human likeness on social presence and social influence. *Computers in Human Behavior*, 124, 106926.
Lury, C. (1996). *Consumer culture*. Cambridge, UK: Polity Press.
Mac Namee, B., Cunningham, P., Byrne, S., & Corrigan, O. I. (2002). The problem of bias in training data in regression problems in medical decision support. *Artificial Intelligence in Medicine*, *24*(1), 51–70.
Mansoury, M., Abdollahpouri, H., Pechenizkiy, M., Mobasher, B., & Burke, R. (2020). Feedback loop and bias amplification in recommender systems. In *Proceedings of the 29th ACM international conference on information & knowledge management* (pp. 2145–2148).
Mohri, M., Rostamizadeh, A., & Talwalkar, A. (2018). *Foundations of machine learning*. Cambridge, MA: MIT Press.
Nyrhinen, J., Sirola, A., Koskelainen, T., Munnukka, J., & Wilska, T. A. (2024). Online antecedents for young consumers' impulse buying behavior. *Computers in Human Behavior*, *153*, 108129.
Olteanu, A., Castillo, C., Diaz, F., & Kıcıman, E. (2019). Social data: Biases, methodological pitfalls, and ethical boundaries. *Frontiers in Big Data*, *2*, 13.
Paavola, L., & Cuthbertson, R. (2021). Algorithms creating paradoxes of power: Explore, exploit, embed, embalm. *Information Systems Management*, *38*(4), 358–371.
Patro, G. K., Porcaro, L., Mitchell, L., Zhang, Q., Zehlike, M., & Garg, N. (2022, June). Fair ranking: A critical review, challenges, and future directions. In *Proceedings of the 2022 ACM conference on fairness, accountability, and transparency* (pp. 1929–1942).
Petty, R. D., Harris, A. M. G., Broaddus, T., & Boyd III, W. M. (2002). Regulating target marketing and other race-related advertising practices. *Michigan Journal of Race & Law*, *8*, 335–394.

Reinikainen, H., Kari, J. T., & Luoma-aho, V. (2020). Generation Z and organizational listening on social media. *Media and Communication*, *8*(2), 185–196.

Rissanen, H., & Luoma-aho, V. (2016). (Un)willing to engage? First look at the engagement types of millennials. *Corporate Communications: An International Journal*, *21*(4), 500–515.

Roselli, D., Matthews, J. & Talagala, N. (2019). Managing bias in AI. In *Proceedings of WWW '19: The Web Conference (WWW '19)*, May 13, 2019, San Francisco, USA. ACM, New York, NY, USA.

Ruckenstein, M., & Granroth, J. (2020). Algorithms, advertising and the intimacy of surveillance. *Journal of Cultural Economy*, *13*(1), 12–24.

Russell, S., & Norvig, P. (2020). *Artificial intelligence: A modern approach*. London: Pearson Education Limited.

Savolainen, I., Oksanen, A., Kaakinen, M., Sirola, A., Zych, I., & Paek, H. J. (2021). The role of online group norms and social identity in youth problem gambling. *Computers in Human Behavior*, *122*, 106828.

Steck, H., Baltrunas, L., Elahi, E., Liang, D., Raimond, Y., & Basilico, J. (2021). Deep learning for recommender systems: A Netflix case study. *The AI Magazine*, *42*(3), 7–18.

Varnali, K. (2021). Online behavioral advertising: An integrative review. *Journal of Marketing Communications*, *27*(1), 93–114.

Willett, R. (2008). Consumer citizens online: Structure, agency, and gender in online participation. In D. Buckingham (Ed.), *Youth, identity, and digital media. The John D. and Catherine T. MacArthur Foundation series on digital media and learning* (pp. 49–70). Cambridge, MA: The MIT Press.

Wilska, T.-A. (2017). Youth and generations in consumption. In M. Keller, B. Halkier, T.-A. Wilska, & M. Truninger (Eds), *Routledge handbook on consumption* (pp. 314–325). Abingdon, UK: Routledge.

Yuan, Y., Wang, F. Y., & Zeng, D. (2017). Competitive analysis of bidding behavior on sponsored search advertising markets. *IEEE Transactions on Computational Social Systems*, *4*(3), 179–190.

4. Careful or carefree? Young people and information non-transparency on social media

Katerina Tsetsura, Vilma Luoma-aho and Eero Rantala

4.1 INTRODUCTION

Today, people rely on social media for different information needs (Ho-Dac, 2020). Information-seeking generally aims to diminish uncertainty (Case & Given, 2016). Previous studies suggest that unclear information leads to distrust and increased uncertainty (Basch et al., 2020). The quality and clarity of information on social media have implications beyond individuals as they affect societies' safety, welfare, health, and even democracy (Di Domenico & Visentin, 2020). Many argue that information inaccuracies and misinformation (that is, information that is false, but not created to cause harm) pollute the information landscape, threaten peace and security, and often disproportionately affect the most vulnerable populations (Tsetsura et al., 2024).

Non-transparency, which refers to the non-disclosure of sources that might have influenced the final text or images that appear in social media or other types of media (Tsetsura & Kruckeberg, 2017), and information inaccuracies, whether intentional or not, are increasingly accepted as the new normal of information in social media, and platform owners continue to debate the responsibilities for policing inaccuracies (Aral, 2020). Social media platforms have proven to be key places for the spread of misinformation, with ramifications spilling over to the offline world (Tandoc et al., 2018; Westerman et al., 2014), and a "contamination of the digital world" (Di Domenico & Visentin, 2020, p. 409), where unverified (through multiple sources by professional journalists) facts and opinions shape individuals' views. Source transparency has in part been replaced by online communities supporting their own sets of beliefs (Braddock & Morrison, 2020). At worst, a hype loop of tailored social media content directly influences individuals' emotions, social relationships, consumption, and actions (Aral, 2020).

Previous studies suggest young people trust information that is shared and generated among users and influencers of specific social media platforms, whereas older generations trust information disseminated by institutions and organizations (Herrando et al., 2019; Reinikainen et al., 2020). This exploratory study examines the attitudes of young people on social media toward non-transparent and unverified information. We ask whether attitudinal differences among young people exist in their acceptance of non-transparent information they receive from social media. Is it possible that young people are less likely to care about the transparency of information?

Understanding young people's attitudes toward the transparency of information is crucial. Transparency is a cornerstone of authentic and trustworthy communication (Tsetsura & Kruckeberg, 2017). Transparency about the origins of information allows citizens to evaluate the perceived quality of information (Ter Mors et al., 2010) and the credibility of information (Westerman et al., 2014) and ultimately contributes to one's acceptance or rejection of information. The complexity of today's information space presents a global challenge, particularly given the abundant online information and its impact on young people's decision-making in a variety of contexts, from commercial to political (Thangavel et al., 2022). A study from Stanford showed that young people have difficulty identifying credible content online (Wineburg et al., 2016). But does credibility matter to young people consuming social media online? Credibility has become an increasingly complex concept. A recent study of teenagers in Brazil, for instance, showed that traditionally credible sources, such as teachers and parents, are only relevant in some circumstances (Almeida et al., 2022).

Traditionally, information transparency has influenced the quality and credibility of information, particularly among communication professionals and adult consumers (Tsetsura & Kruckeberg, 2017). This study surveyed 15- to 29-year-olds in Finland and the UK to examine whether (and if so, how?) information transparency matters to young people in these two countries. Finland, which consistently ranks in the top two most transparent countries in the world (Transparency International, 2021), has recently seen debates on the limits of transparency (Koskela et al., 2021). The UK, ranked in the top 10–15 most transparent countries, has witnessed several social media misinformation and privacy breach scandals in recent years, associated with Cambridge Analytica and Brexit (García-Orosa, 2021).

Little research exists on what factors make young people trust messages on social media even when the information remains non-transparent to some degree. The results of a commercial study on millennials in the UK, for instance, suggested that the credibility of the native advertising message was defined not by whether the message was transparent (i.e., clearly marked as paid) but by whether it was perceived to be *good enough* (Arnold, 2018). Among the first of its kind, in this study, we ask, how do young people perceive useful social media content where transparency may be lacking?

This study offers a more nuanced examination of young people who consume information online. We explored how young people view the non-transparency of information or transparency of information on social media and looked at cultural differences between the attitude profiles in Finland and the UK. Specifically, we focused on determining whether profiles of young people differed between two countries based on trust in video bloggers (hereafter "vloggers"), and sponsored content's influence on buying decisions. The results of this study demonstrate the differences we explored and offer practical implications for addressing the transparency and credibility of messages online among various profiles of young people in two countries.

In what follows, we first discuss information credibility and trust in the context of social media and its consumption by young people. Then, we define factors contributing to non-transparency of online information, present research questions, and discuss the research methodology. Next, we report the findings of our survey and offer a profiling of young people based on the Latent Profile Analysis (LPA). Finally, we discuss the study's findings and limitations and offer implications for future studies.

4.2 THEORETICAL FRAMEWORK AND RESEARCH QUESTIONS

4.2.1 Trust and Information Credibility on Social Media

As vast numbers of internet users generate content, evaluating the truthfulness of information on social media has become increasingly challenging (Tandoc et al., 2018). Social media platforms have altered not only the process of information production but also its consumption, and many individuals rely on social media for information related to their health, safety, and socializing (Bailey & Hsieh-Yee, 2020). The competition for users' time and attention on social media platforms has become an ethic-blurring game. Among the reported challenges for societies of information on social media are verification difficulties, false representations within individual profiles, and even harassment and bullying (Bailey & Hsieh-Yee, 2020).

Truth as an ideal of information, builds on the ideal of "objectivity with complete transparency" by gathering and disseminating accurate, completed, fact-based, unbiased information (Tsetsura & Kruckeberg, 2017, p. 4). Although much social media content may appear to be true, it most often does not follow the ethical guidelines associated with truthful content creation and fact-checking (Tandoc et al., 2018).

Untrue social media content has been labeled as problematic information, misinformation, or even disinformation. Problematic information is not just distorted, misleading, and inaccurate, but also contains "inappropriately attributed, or altogether fabricated information" (Jack, 2017, p. 1096). As social media continues to gain popularity among users, some have suggested that we are facing an epidemic of generalized scepticism (Fletcher & Nielsen, 2019), where online information is approached with a distrustful stance.

The plasticity of digital information and the hyperlinked structure of social media challenge trust and credibility evaluations, as modifications, misrepresentations, and contextualization can be easily adapted to information on social media. Trust is seen as a mediator that determines users' and consumers' attitudes toward other people, objects, and events (Yang & Lin, 2014). As Kelton et al. (2008) suggest, trust is always context-specific – it cannot be generalized to everything online. As Kelton et al. (2008) argued, "the propensity to trust information is a stable disposition and exerts a positive influence on the likelihood and level of trust in a specific source" (p. 373). In the online environment, trust can also be associated with a specific community online (Hsu, 2017).

Previous research shows that young people's information adoption relies on information usefulness and trust, and cues are taken from several aspects, ranging from whether the information appears complete to how it is presented online (Mou et al., 2016). Trust in social media content is often seen as a combination of evaluations of the target's benevolence, integrity, and competency (Warner-Søderholm et al., 2018). As trust in social media is a combination of trust in the source and the technology platform used (Warner-Søderholm et al., 2018; Yang & Lin, 2014), understanding social media trust is a complex process.

When it comes to generations, inter-generational differences have been reported concerning trust. Broadly speaking, individuals who were born between the mid to late 1990s and 2012 have been reported to trust information that is shared and generated among users of specific services, whereas the older generations rely more on information generated by institutions and organizations (Herrando et al., 2019). Trust and experiences are related, and information that can meet the users' taste requirements or elicit a positive emotional response is more trusted (Kelton et al., 2008). However, little research has been done to look at intra-generational differences in trust. This exploratory study, the first of its kind, aims to look at differences among young people between 15 and 29 years old in two countries: Finland and the UK.

4.2.2 Usefulness of Information and Credibility

Information usefulness, as it relates to credibility and its implications for transparent communication, has been defined as the extent to which received information is perceived as valuable (Luo et al., 2018), and its usefulness is determined by the adoption of information (Shen et al., 2013). Credibility is understood as a combination of the credibility evaluations of a message and its source (Metzger & Flanagin, 2013). Information may be processed more quickly based not on the message itself but on the perceived credibility of a source (Chaoguang et al., 2018). On social media, credibility evaluations are aided by access to the content author and their role in the community in question: much of social media content is created and shared via open social media profiles of individuals and organizations both authentic and fake (Wierzbicki, 2018).

Source credibility relies on the trustworthiness and expertise of the source (Van Boekel et al., 2017). Online credibility remains easily manipulated and relies on the transparency signaling of the individuals contributing to the content (Bailey & Hsieh-Yee, 2020). Overall, there seems to be agreement that the same sources online are less credible than their offline equivalents (see, for example, Jo, 2005), perhaps owing to limited fact-checking possibilities because of the volume of information available online (Liu et al., 2015).

The question addressed in this study is whether perceived transparency of online information is a degree of variance among young people's perceptions. Do all young people equally care about transparency and credibility indicators? Are there differences in the way young people perceive the usefulness of information? What, if any, differences exist among young people from different countries? To answer these questions, we conducted the first-of-its-kind exploratory survey to examine (RQ1): *What kind of attitude profiles toward the non-transparency of social media content emerge among young people in Finland and the UK?*

4.2.3 Two Countries in Focus: Social Media Use in Finland and the UK

Social media use among young people has been described as a phenomenon of the information society, and many Western democracies have high numbers of young people who utilize social media services. Both in the UK and in Finland, press freedom is among the highest in the world (RSF, 2022). There are no remarkable differences in digital media usage between Finland and the UK. The European Social Survey (2022) reports both countries as having very active internet users, and the social media penetration rate in both is between 60 and 70 percent.

Differences emerge when it comes to trust in society: Finland leads with higher reported amounts of generalized trust as well as higher trust in institutions (European Social Survey, 2022). On the other hand, regarding consumption and social media use, the UK leads the way with a longer history of a consumer society (Wilska & Pedrozo, 2007). Although Finland was a pioneer country in terms of early digitalization and mobile communication (e.g., Wilska & Pedrozo, 2007), its commercialization of digital content has been more recent than in the UK (Merisavo et al., 2007).

Finland has been ranked as the second most transparent country in the world, and the UK has been ranked number 11 (Transparency International, 2021). This would suggest that attitudes toward information non-transparency in Finland may be slightly more critical than in the UK. The UK has witnessed several social media misinformation and privacy breach scandals in recent years, with Cambridge Analytica and the Brexit campaign as proven examples, suggesting that people in the UK are more accustomed to breaches of privacy. Taking these developments and similarities as our background, we examine (RQ2): *Are there differences in attitudes toward information in social media between profiles of young people in Finland and the UK?*

We particularly explored how young people view the non-transparency of information on social media and looked at cultural differences between the attitude profiles in Finland and the UK. Specifically, we focused on determining whether profiles of young people from the two countries differed based on specific theory-driven variables: belonging to a social media community, trust in vloggers, and sponsored content's influence on buying decisions.

4.2.4 Factors that Influence Non-transparency

Based on the previous studies, we identified three factors that might influence the perceived transparency of online information: (1) belonging to a social media community, (2) trust in vloggers, and (3) sponsored content's influence on buying decisions. In this section, we discuss each of these three factors.

First, belonging to a social media community can be a factor in whether young people perceive information as transparent and authentic. Young people are generally more susceptible to peer influence and more inclined toward social comparison (e.g., Stok et al., 2016), and social media intensifies these (e.g., Tuominen et al., 2022). In addition, unrealistic comparisons with peers may shape the identity formation process of young people for better or worse (Valkenburg et al., 2021). The social media communities young people belong to shape their actions (Cooley & Parks-Yancy, 2019), as a sense of community may form quickly in computer-mediated spaces during interaction and sharing (Lim, 2014). Implications can vary, as emotions and information shared online in social media communities can spread beyond their original context (Bowden et al., 2018).

As social media environments become increasingly addictive, simultaneous excessive targeting of information and advertising decreases the visibility of novel perspectives and emphasizes in-group cohesion (Lewandowsky & van der Linden, 2021). As algorithms curate content on social media platforms and in online communities, a closed system of ideological asymmetry may form. Polarized information and strong emotional reactions may skew perceptions of reality (Aral, 2020; Di Domenico & Visentin, 2020).

Second, trust in vloggers is increasingly a factor in authenticity of communication. Social media communities have expanded rapidly, and young people are exposed to both advertising and more indirect commercial communication from a variety of sources (Reinikainen et al., 2020). Social media content that is useful to young people creates loyalty (Reinikainen et al., 2020), and studies show that many social media influencers are highly successful in building trust among their online communities (Marôpo et al., 2020). Such illusionary high-trust friendships or parasocial relationships are positively linked with follower purchase intentions and decisions (Djafarova & Rushworth, 2017). In particular, vloggers have become major influencers on young people's consumption of information and buying decisions (Chen et al., 2022).

Third, sponsored content's influence on buying decisions can also shed light on how young people perceive information to be transparent and credible (Devedi et al., 2017; Nusairat et al., 2021). Specifically, prior research demonstrated clearly that online sponsored content influences perceived authenticity and affects buying decisions among young people (Luoma-aho et al., 2019).

Based on this discussion, we examine (RQ3): *How can profiles be characterized by the three descriptive factors of belonging to a social media community, trust in vloggers, and sponsored content's influence on buying decisions?*

4.3 METHODOLOGY

4.3.1 Data and Measures

The data (n = 2,558) was collected by surveying 15–29-year-olds in 2019 in Finland (n = 1,281) and in the UK (n = 1,277). The respondents were recruited using a random sampling from an online research panel and the sample was representative of age, gender, and area of residence. We utilized contextual single-item variables measured on a 5-point Likert-type scale (1 = agree, 5 = disagree or 1 = never, 5 = always), because the variables had clear interpretations, and in such cases, single-item measures are generally preferred (Ang & Eisend, 2018). Descriptive statistics of the indicator variables used in the profile modeling are presented in Table 4.1, and descriptive statistics of the covariates are shown in Table 4.2.

4.3.2 Methodological Approach

To examine the consumer profiles and their differences between two countries, we conducted LPA (Latent Profile Analysis) through a multigroup procedure with MPlus v. 7 software. LPA was chosen because it is a flexible mixed-model method that provides information about different subpopulations based on their attitudes (Morin et al., 2016). A multigroup modeling procedure of LPA also shows whether the profiles and their number are the same between countries. The maximum likelihood robust estimator was used for handling missing data and non-normality issues.

4.4 RESULTS

4.4.1 Test of Similarity between Profiles in Finland and the UK

The similarity between the profiles across Britain and Finland was first assessed, and the above-described procedure by Morin et al. (2016) was applied. LPA was performed separately for both country samples to test their configural similarity, that is, to determine whether there were the same number of profiles in each country. The BIC (Bayesian Information Criterion), sample-size-adjusted BIC (SABIC), and corrected version of the Akaike information criterion (CAIC) (Morin et al., 2016) were used as the main indicators for our decision regarding the optimal number of profiles (lower values indicate a better model). Additionally, the Lo-Mendell-Rubin (LMR) adjusted test and bootstrapped likelihood ratio test were used. Alongside numerical information and criteria, interpretability of the profiles was considered when choosing the profile structure (see Morin et al., 2016).

Four was regarded as the optimal number of profiles for both countries because BIC, SABIC, and CAIC indicated no improvement if more profiles were added (see Statistical Appendix 1 and Figures 4A.1 and 4A.2). In addition, the entropy level (> 0.8) and the classification probability (> 0.7) were at satisfactory levels according to Nagin (2005). The LMR test was consistent with the information criteria in Finland ($p = 0.57$, 4 vs 5 profiles). However, in the UK sample, the LMR supported the two-profile solution ($p = 0.060$, 2 vs 3 profiles) and did not agree with the information criteria. Regardless of that, we chose a four-profile solution for both countries, because it provided meaningful interpretations.

Next, we tested structural similarity by comparing a freely estimated configural model with a structural model in which the means of the indicators are

Table 4.1 *Descriptive statistics of the attitudes toward transparency*

Indicators	Mean	Variance	*N**
As long as the content is useful to me...(1 = totally disagree, 5 = totally agree)	(UK/Finland/ Overall)	(UK/Finland/ Overall)	(UK/Finland/ Overall)
It does not matter who wrote it	3.061/2.912/2.988	1.786/1.698/1.748	1192/1161/2353
It does not matter if it contains sponsored content	3.071/2.897/2.985	1.587/1.564/1.583	1189/1170/2359
It does not matter if I do not know the source of the information	2.848/2.459/2.655	1.661/1.592/1.664	1184/1167/2351
It does not matter if the content is true or not	2.457/2.222/2.340	2.177/1.668/1.938	1186/1169/2355
It does not matter to me if my information is stored by the search engine	3.038/2.642/2.841	1.765/1.712/1.778	1171/1159/2330
It does not matter that search engines record my conversations	2.600/2.096/2.347	2.041/1.928/2.047	1164/1169/2333

Note: *Max *N*: UK = 1,277; Finland = 1,281; Overall = 2,558.

Table 4.2 *Descriptive statistics of the covariates*

Covariates	Mean* (UK/Finland/ Overall)	Variance (UK/Finland/ Overall)	*N*** (UK/Finland/ Overall)
Belonging to a social media community (1= totally disagree, 5 = totally agree)	3.731/3.454/3.594	1.408/1.523/1.484	1212/1189/2401
Trust vloggers (1 = Never, 5 = Always)	3.253/2.944/3.101	1.350/1.284/1.341	1187/1147/2334
Sponsored content's influence buying decisions (1 = Never, 5 = Always)	3.031/2.861/2.948	1.598/1.475/1.545	1180/1138/2318

Note: *The mean in the case of the dichotomous variable gives the proportion of how many responded to option one of the variable; **Max *N*: UK = 1,277; Finland = 1,281; Overall = 2,558.

constrained in an equivalent way to the profiles between countries (see Morin et al., 2016). BIC, SABIC and CAIC indexes showed that a more restricted model was not supported. This is because the values for these indexes were higher than in freely estimated model.[1] The solution indicated that the structures of profiles between countries were different. Therefore, we conducted a separate analysis for Finland and the UK. See Tables 4.3a and b for the results of the first phase of these analysis.

4.4.2 What Kind of Profiles Emerge in Finland and the UK (RQ1) and How Do They Differ (RQ2)?

Even though the structural test did not support the exact similarity of the profiles, the two countries' profiles were very similar regarding their contents. In both countries, three of the four profiles were somewhat identical: one profile

Table 4.3a *Profiles from the configural modeling: the UK*

-	Means of Profiles PF. 1	Means of Profiles PF. 2	Means of Profiles PF. 3	Means of Profiles PF. 4	Variances PF. 1–4
It does not matter who wrote it	2.008	2.936	3.774	4.473	0.937
It does not matter if it contains sponsored content	2.127	3.119	3.454	4.388	0.935
It does not matter if I do not know the source of the information	1.599	3.148	3.196	4.425	0.624
It does not matter if the content is true or not	1.182	3.476	1.406	4.594	0.29
It does not matter to me if my information is stored by the search engine	2.055	3.418	3.028	4.359	1.075
It does not matter that search engines record my conversations	1.537	3.204	2.175	4.287	0.996

Note: All means and variances differ significantly ($p < 0.05$); variances are the same for all profiles in each country because of variance constraints.

Table 4.3b Profiles from the configural modeling: Finland

-	Means of Profiles PF. 1	Means of Profiles PF. 2	Means of Profiles PF. 3	Means of Profiles PF. 4	Variances PF. 1–4
It does not matter who wrote it	2.939	3.578	2.253	4.304	1.179
It does not matter if it contains sponsored content	3.081	3.618	2.137	4.303	0.953
It does not matter if I do not know the source of the information	2.85	3.065	1.533	4.364	0.66
It does not matter if the content is true or not	2.879	2.271	1.33	4.37	0.652
It does not matter to me if my information is stored by the search engine	3.246	2.83	1.827	4.407	0.951
It does not matter that search engines record my conversations	3.403	1.267	1.157	4.623	0.242

Note: All means and variances differ significantly ($p < 0.05$); variances are the same for all profiles in each country because of variance constraints.

(UK profile 1/Finland profile 3) was more critical of non-transparency than the others (the lowest means indicated disagreement). We named these profiles *the Careful*. In both countries, there were also profiles (UK profile 2/Finland profile 1) that had means somewhat near the midpoint of the scale for every indicator. We named these profiles *the Neutral*. In both countries, there was a profile that was quite ignorant about non-transparency (UK profile 4/Finland profile 4), and we called these profiles *the Carefree*.

The differences between the two countries occurred in the last profiles near the neutral profile (UK profile 3/Finland profile 2), which differed in terms of the two statements only. In this profile, it was more important for the UK respondents to receive truthful information than for the Finnish respondents. We called UK profile 3 *the Truth Seekers*, whereas in profile 3 in Finland, the respondents were more critical than in the UK about having their conversations recorded by search engines. We labeled the Finland profile 3 *the*

Private. Figures 4.1 and 4.2 show the similarity and dissimilarity in the profiles between the countries and reveal the sizes of the profiles.

With profile sizes, approximate comparisons can be made. Despite the configural modeling suggesting some differences between Finland and the UK, three of the four profiles were similar for both countries.

The careful profile was the biggest in both countries, with a large difference between countries (Finland 47 percent and the UK 35 percent). The neutral and the private/truth seekers were quite similar in size. The smallest profile in both countries was the carefree (Finland 12 percent and the UK 18 percent). Almost one-fifth of young people do not care about transparency on social media in the UK, whereas this fraction is closer to one-tenth in Finland. Overall, it seems, based on these approximate comparisons, that Finnish young people care more about transparency on social media compared with young people in the UK (large size differences between the careful and the carefree profiles).

4.4.3 How Can the Profiles be Characterized (RQ3)?

To answer RQ3, we predicted means of covariates with distal outcome modeling for variables measuring "feeling of belonging to a social media community," "trust in vloggers," and "how much social media has influenced their buying decisions." The results are presented in Figure 4.3 for the UK and

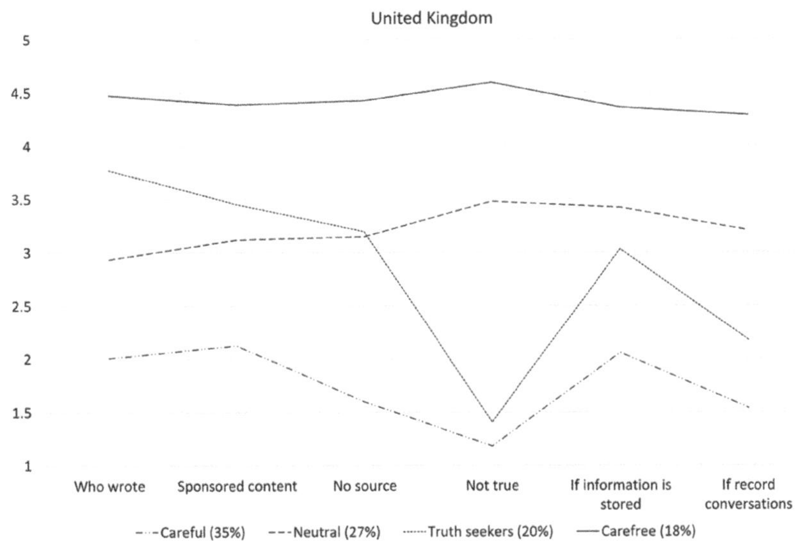

Figure 4.1 *Profile plot for the UK*

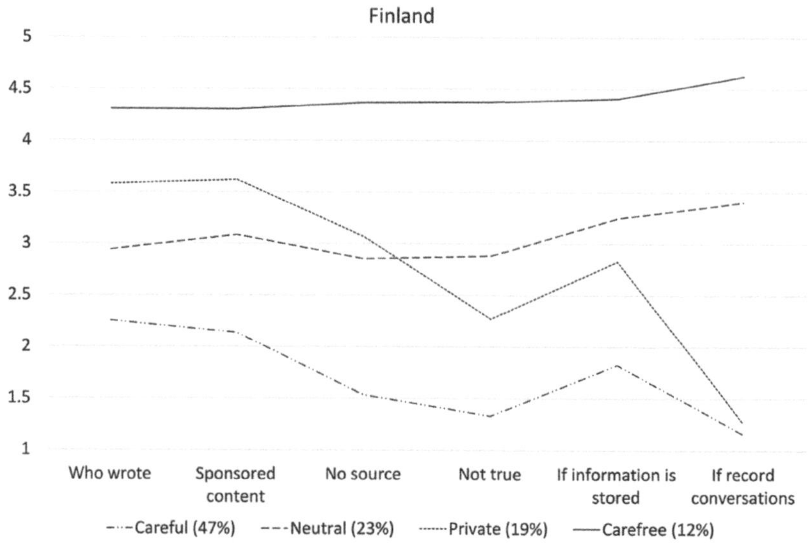

Figure 4.2 Profile plot for Finland

Figure 4.4 for Finland with specific mean values. See Statistical Appendix 2 for test statistics for the mean differences in both countries (Tables 4A.1a–4A.1c).

Overall, the results demonstrated statistically significant differences between the truth-seekers and carefree profiles for the UK and private and carefree profiles in Finland across all categories. We also observed differences between careful and carefree profiles in both counties across all categories. Trust in vloggers was lowest for careful profiles in each country, the UK and Finland, and such trust was highest for carefree profiles in each country.

4.5 DISCUSSION

This chapter contributes to our overall understanding of how young people perceive online information. We are among the first to demonstrate that young people are not homogeneous consumers of information on social media and, overall, are *more careful than carefree*, in contrast with previous research (Herrando et al., 2019). From a conceptual standpoint, this study considers the different needs and expectations of different profiles of young people, whether young people are concerned with privacy issues or whether they strive to distinguish unauthentic content from authentic. The study, hence, offers a

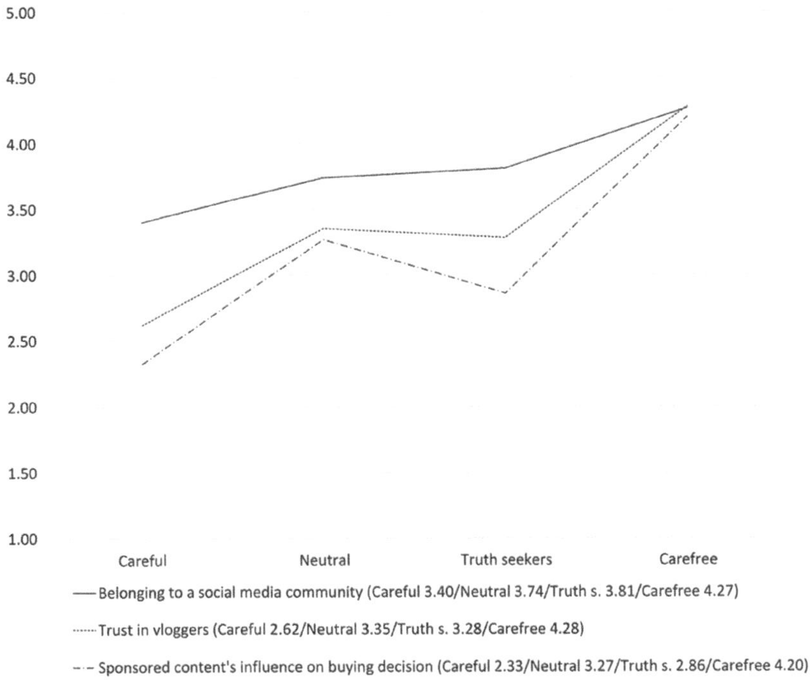

Figure 4.3 *Means of the covariates for the UK*

conceptually new way to understand the differences among young people in terms of their perception of online information transparency and credibility.

These results are positive on several levels: most young people still seem to be critical of online information non-transparency, or at least adopt a neutral attitude toward it in both Finland and the UK. In both countries, the *carefree* profile was the smallest one; however, a need for media literacy efforts and education development in both countries still exists. Even though the carefree represents a small group in this study, as membership of social media communities plays a major role for them, it is possible that their carefree attitude can spill over to others, as young people are highly motivated and influenced by their peers and influencers. Critical questions that still need to be answered in the future are how many of the social media influencers that impact most young people belong to the carefree group and whether they are careful of their influence related to attitudes about non-transparency.

The acknowledged problems related to non-transparency acceptance are, hence, those of only a small minority of young people. Why this group is

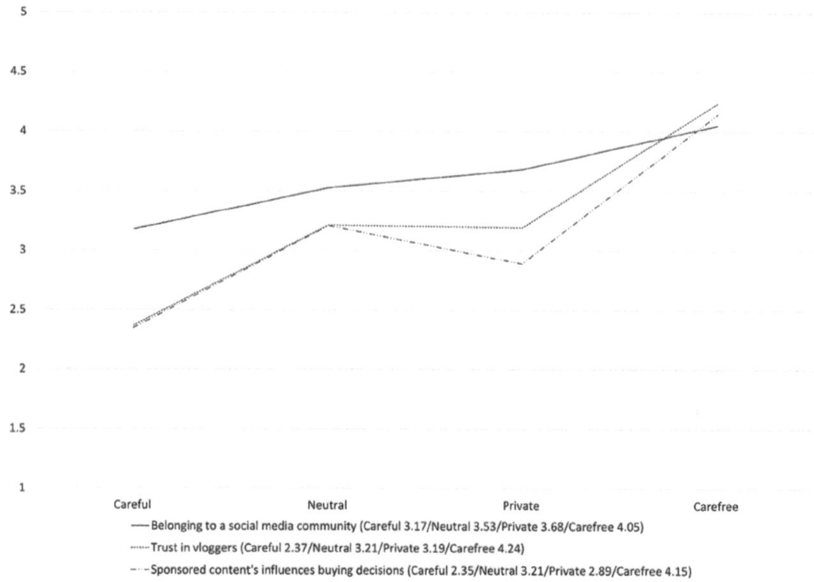

Figure 4.4 *Means of the covariates for Finland*

somewhat larger in the UK than in Finland can perhaps, in part, be explained by the more mature stage of consumerism in society or the higher level of overall education in Finland. The transparency of the societies can also play a role, as Finland is one of the most transparent societies in the world.

In short, this research supported prior research about young people's perceptions of transparency of online information. Four distinct profiles of young people in both Finland and the UK were identified, based on the three factors: belonging to a social media community, trust in vloggers, and sponsored content's influence on buying decisions. Young people perceive information differently depending on their attitudes toward each of these factors. This exploratory study showed that young people are not homogeneous consumers of online information.

4.5.1 Practical Implications

The study has several practical implications. First, brands and organizations need to remember that the data they collect and use about young people may be skewed to include more of the carefree, who are willing to share information to gain benefits. Those people, however, might not represent the entire population, as the majority in both countries remain careful and critical of non-transparency.

New questions and responsibilities emerge for public sector and civic society organizations and for companies and advertisers: Should these carefree young people be better protected on social media? How should we ensure that this group, which is so ignorant of its social media data traces, is not taken advantage of? If a carefree approach enables the spread of non-verified information and misinformation in society, should the carefree be the new focus of societal resilience building? And more precisely, how can the social media communities that they place their trust in best be reached?

4.5.2 Avenues for Future Research

There are some limitations to be noted. First, the survey instrument has utilized an original construction of non-transparency that has not been tested in previous studies. Second, future studies focusing on non-transparency should focus on further scale development and model testing for social media information transparency.

Third, two North European countries were included in this data collection. Thus, future studies could compare countries that have even more distinctive cultural differences. For instance, understanding how young people in non-Western (e.g., China) and hybrid Western/non-Western countries (e.g., Kazakhstan) perceive trust and transparency online would further propel our knowledge of how non-transparent information on social media is perceived and understood. Fourth, although strict similarity was lacking within the profiles between the countries, three of the four profiles were approximately identical when viewed, meaning that the results indicate the proposed distinctions in profiles of young people, as related to attitude profiles toward the non-transparency of social media in terms of usefulness of information and credibility, might be global. This similarity could be examined in future research.

AUTHORS' NOTE

The work of Eero Rantala was supported by the Strategic Research Council (SRC) at the Academy of Finland (#352544).

NOTE

1. BIC: 44523.62–44639.98; SABIC: 44310.75–44503.36; CAIC: 44590.62–44 682.98.

REFERENCES

Almeida, C., Macedo-Rouet, M., de Carvalho, V. B., Castilhos, W., Ramalho, M., Amorim, L., & Massarani, L. (2022). When does credibility matter? The assessment of information sources in teenagers' navigation regimes. *Journal of Librarianship and Information Science, 55*(1), 218–231. https://doi.org/10.1177/09610006211064647.

Ang, L., & Eisend, M. (2018). Single versus multiple measurement of attitudes. *Journal of Advertising Research, 58*(2), 218–227. doi:10.2501/Jar-2017-001.

Aral, S. (2020). *The hype machine: How social media disrupts our elections, our economy, and our health – and how we must adapt.* Sydney, Australia: Currency.

Arnold, A. (2018, Jan 21). Millennials hate ads but 58% of them wouldn't mind if it's from their favorite digital stars. *Forbes.* https://www.forbes.com/sites/andrewarnold/2018/01/21/millennials-hate-ads-but-58-of-them-wouldnt-mind-if-its-from-their-favorite-digital-stars/.

Bailey, T. C., & Hsieh-Yee, I. (2020). Combating the sharing of false information: History, framework, and literacy strategies. *Internet Reference Services Quarterly, 24*(1–2), 9–30. https://doi.org/10.1080/10875301.2020.1863286.

Basch, C. H., Mohlman, J., Hillyer, G. C., & Garcia, P. (2020). Public health communication in time of crisis: Readability of on-line COVID-19 information. *Disaster Medicine and Public Health Preparedness, 14*(5), 635–637.

Bowden, J. L. H., Conduit, J., Hollebeek, L. D., Luoma-aho, V., & Solem, B. A. A. (2018). The role of social capital in shaping consumer engagement within online brand communities. In K. Johnston & M. Taylor (Eds), *The handbook of communication engagement* (pp. 491–504). Hoboken, NJ: Wiley Blackwell.

Braddock, K., & Morrison, J. F. (2020). Cultivating trust and perceptions of source credibility in online counternarratives intended to reduce support for terrorism. *Studies in Conflict & Terrorism, 43*, 468–492.

Case, D. O., & Given, L. M. (2016). *Looking for information: A survey of research on information seeking, needs, and behavior* (4th edition). Bingley, UK: Emerald.

Chaoguang, H., Feicheng, M., Yifei, Q., & Yuchao, W. (2018). Exploring the determinants of health knowledge adoption in social media: An intention-behavior-gap perspective. *Information Development, 34*, 346–363. doi:10.1177/0266666917700231.

Chen, Y., Tingchi Liu, M., Liu, Y., Chang, A. W. Y., & Yen, J. (2022). The influence of trust and relationship commitment to vloggers on viewers' purchase intention. *Asia Pacific Journal of Marketing and Logistics, 34*(2), 249–267.

Cooley, D., & Parks-Yancy, R. (2019). The effect of social media on perceived information credibility and decision making. *Journal of Internet Commerce, 18*, 249–269.

Devedi, P., Sujatha, R., & Pathak, R. (2017). A study on parameters of online reviews content that influence consumers buying behaviour: An Indian perspective. *Journal of Business and Retail Management Research (JBRMR), 11.* https://jbrmr.com/cdn/article_file/content_87473_17-07-22-07-32-40.pdf.

Di Domenico, G., & Visentin, M. (2020). Fake news or true lies? Reflections about problematic contents in marketing. *International Journal of Market Research, 62*, 409–417.

Djafarova, E., & Rushworth, C., (2017). Exploring the credibility of online celebrities' Instagram profiles in influencing the purchase decisions of young female users. *Computers in Human Behavior, 68*, 1–7.

European Social Survey (2022). *Exploring public attitudes, informing public policy. Selected findings from the first nine rounds.* Report 4. European Social Survey 2022.

Fletcher, R., & Nielsen, R. (2019). Generalized scepticism: How people navigate news on social media. *Information Communication and Society*, *22*, 1–19.

García-Orosa, B. (2021). Disinformation, social media, bots, and astroturfing: The fourth wave of digital democracy. *Profesional de la información*, *30*(6), e300603. https://doi.org/10.3145/epi.2021.nov.03.

Herrando, C., Jimenez-Martinez, J., & Martin-De Hoyos, M. J. (2019). Tell me your age and I tell you what you trust: The moderating effect of generations. *Internet Research*, *29*, 799–817. doi:10.1108/IntR-03-2017–0135.

Ho-Dac, N. (2020). The value of online user generated content in product development. *Journal of Business Research*, *112*, 136–146.

Hsu, L. (2017). Investigating community members' purchase intention on Facebook fan page: From a dualistic perspective of trust relationships. *Industrial Management + Data Systems*, *117*, 766–800. doi:10.1108/IMDS-05-2016–0180.

Jack, C. (2017). Lexicon of lies: Terms for problematic information. *Data & Society*, *3*, 1094–1096.

Jo, S. (2005). The effect of online media credibility on trust relationships. *Journal of Website Promotion*, *1*, 57–78.

Kelton, K., Fleischmann, K. R., & Wallace, W. A. (2008). Trust in digital information. *Journal of the American Society for Information Science*, *59*, 363–374. doi:10.1002/asi.20722.

Koskela, M., Enell-Nilsson, M., & Hjerppe, C. (2021). Communication practitioners' views on transparency and its impact in Finnish companies. In H. Katajamäki, M. Enell-Nilsson, H. Kauppinen-Räisänen, L. Kääntä, & H. Salovaara (Eds), *Workplace communication IV: Vakki-symposiumi XLI* (pp. 101–116). Vaasa, Finland: VAKKI publications. https://vakki.net/wp-content/uploads/2021/12/WCIV2021A9.pdf.

Lewandowsky, S., & van der Linden, S. (2021). Countering misinformation and fake news through inoculation and prebunking. *European Review of Social Psychology*, *32*, 348–384.

Lim, W. M. (2014). Sense of virtual community and perceived critical mass in online group buying. *Journal of Strategic Marketing*, *22*, 268–283.

Liu, X., Nielek, R., Adamska, P., Wierzbicki, A., & Aberer, K. (2015). Towards a highly effective and robust web credibility evaluation system. *Decision Support Systems*, *79*, 99–108.

Luo, C., Luo, X. R., & Bose, R. (2018). Information usefulness in online third-party forums. *Computers in Human Behavior*, *85*, 61–73. doi:10.1016/j.chb.2018.02.041.

Luoma-aho, V., Pirttimaki, T., Maity, D., Munnukka, J., & Reinikainen, H. (2019). Primed authenticity: How priming impacts authenticity perception of social media influencers. *International Journal of Strategic Communication*, *13*, 352–365.

Marôpo, L., Jorge, A., & Tomaz, R. (2020). "I felt like I was really talking to you!": Intimacy and trust among teen vloggers and followers in Portugal and Brazil. *Journal of Children & Media*, *14*, 22–37. doi:10.1080/17482798.2019.1699589.

Merisavo, M., Kajalo, S., Karjaluoto, H., Virtanen, V., Salmenkivi, S., Raulas, M., & Leppäniemi, M. (2007). An empirical study of the drivers of consumer acceptance of mobile advertising. *Journal of Interactive Advertising*, *7*, 41–50. doi:10.1080/15252019.2007.10722130.

Metzger, M. J., & Flanagin, A. J. (2013). Credibility and trust of information in online environments: The use of cognitive heuristics. *Journal of Pragmatics*, *59*, 210–220.

Morin, A. J. S., Meyer, J. P., Creusier, J., & Biétry, F. (2016). Multiple-group analysis of similarity in latent profile solutions. *Organizational Research Methods*, *19*, 231–254. https://doi.org/10.1177/1094428115621148.

Mou, J., Shin, D. H., & Cohen, J. (2016). Understanding trust and perceived usefulness in the consumer acceptance of an e-service: A longitudinal investigation. *Behaviour & Information Technology*, *36*(2), 125–139. https://doi.org/10.1080/0144929X.2016.1203024.

Nagin, D. S. (2005). *Group-based modeling of development*. Cambridge, MA: Harvard University Press.

Nusairat, N. M., Alroale, M. A., Qeed, M. A., Al-Gasawneh, J. A., & Hammouri, Q. (2021). User-generated content-consumer buying intentions nexus: The mediating role of brand image. *Academy of Strategic Management Journal*, *20*, 1–12.

Reinikainen, H., Munnukka, J., Maity, D., & Luoma-aho, V. (2020). "You really are a great big sister": Parasocial relationships, credibility, and the moderating role of audience comments in influencer marketing. *Journal of Marketing Management*, *36*(3–4), 279–298.

RSF (2022). *2022 World Press Freedom Index*. https://rsf.org/en/rsf-s-2022-world-press-freedom-index-new-era-polarisation.

Shen, X.-L., Cheung, C. M. K., & Lee, M. K. O. (2013). Trust in students' information adoption from Wikipedia. *British Journal of Educational Technology*, *44*, 502–517. doi:10.1111/j.1467–8535.2012.01335.x.

Stok, F. M., de Vet, E., de Ridder, D. T. D., & de Wit, J. B. F. (2016). The potential of peer social norms to shape food intake in adolescents and young adults: A systematic review of effects and moderators. *Health Psychology Review*, *10*, 326–340.

Tandoc, E., Ling, R., Westlund, O., Duffy, A., Goh, D., & Wei, L. (2018). Audiences' acts of authentication in the age of fake news: A conceptual framework. *New Media & Society*, *20*, 2745–2763.

Ter Mors, E., Weenig, M. W. H., Ellemers, N., & Daamen, D. D. L. (2010). Effective communication about complex environmental issues: Perceived quality of information about carbon dioxide capture and storage (CCS) depends on stakeholder collaboration. *Journal of Environmental Psychology*, *30*, 347–357.

Thangavel, P., Pathak, P., & Chandra, B. (2022). Consumer decision-making style of Gen Z: A Generational cohort analysis. *Global Business Review*, *23*(3), 710–728. https://doi.org/10.1177/0972150919880128.

Transparency International (2021). *Corruption perception index 2021*. Report. https://www.transparency.org/en/cpi/2021.

Tsetsura, K., & Kruckeberg, D. (2017). *Transparency, public relations and the mass media: Combating the hidden influences in news coverage worldwide*. New York: Taylor & Francis.

Tsetsura, K., Ngondo, P., & Kiambi, D. (2024, May). *Strategic communication in fact-checking initiatives: Problematics of countering disinformation efforts in Kenya*. Competitive paper presented at the 2024 ICA PRD pre-conference, Brisbane, Australia.

Tuominen, J., Rantala, E., Reinikainen, H., Luoma-aho, V., & Wilska, T.-A. (2022). The brighter side of materialism: Managing impressions on social media for higher social capital. *Poetics*, *92*, 101651.

Valkenburg, P., Beyens, I., Pouwels, J. L., van Driel, I. I., & Keijsers, L. (2021). Social media use and adolescents' self-esteem: Heading for a person-specific media effects paradigm. *Journal of Communication*, *71*, 56–78.

Van Boekel, M., Lassonde, K. A., O'Brien, E. J., & Kendeou, P. (2017). Source credibility and the processing of refutation texts. *Memory & Cognition*, *45*, 168–181.

Warner-Søderholm, G., Bertsch, A., Sawe, E., Lee, D., Wolfe, T., Meyer, J., Engel, J., & Fatilua, U. N. (2018). Who trusts social media? *Computers in Human Behavior*, *81*, 303–315. doi:10.1016/j.chb.2017.12.026.

Westerman, D., Spence, P. R., & Van Der Heide, B. (2014). Social media as information source: Recency of updates and credibility of information. *Journal of Computer-Mediated Communication*, *19*, 171–183.

Wierzbicki, A. (2018). *Web content credibility*. Berlin: Springer. https://doi.org/10.1007/978-3-319-77794-8.

Wilska, T.-A., & Pedrozo, S. (2007). New technology and young people's consumer identities. A comparative study of Finland and Brazil. *Young: Nordic Journal of Youth Studies*, *15*, 343–368. doi:10.1177/110330880701500402.

Wineburg, S., McGrew, S., Breakstone, J., & Ortega, T. (2016). *Evaluating information: The cornerstone of civic online reasoning*. Research report. Stanford Digital Repository. http://purl.stanford.edu/fv751yt5934.

Yang, H.-L., & Lin, C.-L. (2014). Why do people stick to Facebook web site? A value theory-based view. *Information Technology & People*, *27*, 21–37. https://doi.org/10.1108/ITP-11-2012-0130.

APPENDIX 1: ELBOW PLOTS FOR THE UK AND FINLAND

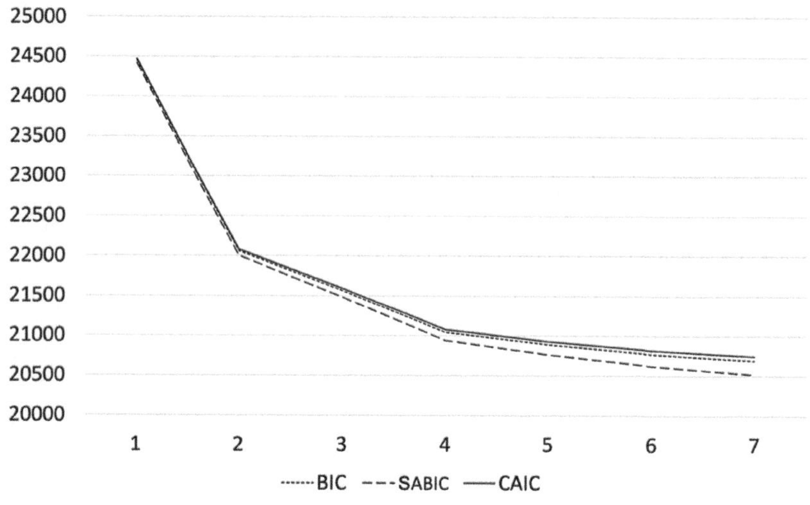

Figure 4A.1 *Elbow plot for the UK*

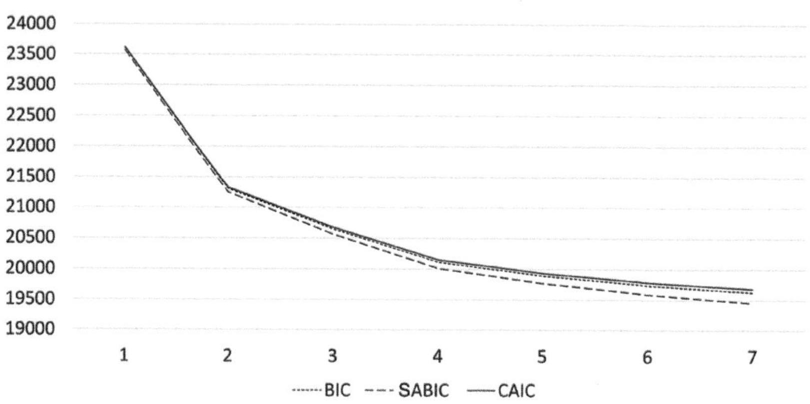

Figure 4A.2 *Elbow plot for Finland*

APPENDIX 2: TEST STATISTICS FOR THE MEAN DIFFERENCES IN THE UK AND FINLAND

Table 4A.1a *Test statistics of mean differences: belonging to a social media community*

Metrics	UK Chi-square	UK *P*-value	Metrics	Finland Chi-square	Finland *P*-value
Overall test	79.001	0.000	Overall test	67.433	0.000
Careful vs Truth Seekers	11.958	0.001	Careful vs Private	19.208	0.000
Neutral vs Truth Seekers	0.373	0.541	Neutral vs Private	1.683	0.195
Truth Seekers vs Carefree	15.405	0.000	Private vs Carefree	8.032	0.005
Careful vs Neutral	13.247	0.000	Careful vs Neutral	13.596	0.000
Careful vs Carefree	78.764	0.000	Careful vs Carefree	63.027	0.000
Neutral vs Carefree	25.430	0.000	Neutral vs Carefree	17.822	0.000

Table 4A.1b *Test statistics of mean differences: trust in vloggers*

Metrics	UK Chi-square	UK *P*-value	Metrics	Finland Chi-square	Finland *P*-value
Overall test	349.094	0.000	Overall test	380.850	0.000
Careful vs Truth Seekers	41.861	0.000	Careful vs Private	77.679	0.000
Neutral vs Truth Seekers	0.489	0.484	Neutral vs Private	0.043	0.835
Truth Seekers vs Carefree	90.037	0.000	Private vs Carefree	76.119	0.000
Careful vs Neutral	76.996	0.000	Careful vs Neutral	121.760	0.000
Careful vs Carefree	339.829	0.000	Careful vs Carefree	322.608	0.000
Neutral vs Carefree	88.653	0.000	Neutral vs Carefree	79.277	0.000

Table 4A.1c *Test statistics of mean differences: sponsored content's influence on buying decisions*

Metrics	UK Chi-square	UK *P*-value	Metrics	Finland Chi-square	Finland *P*-value
Overall test	404.867	0.000	Overall test	333.921	0.000
Careful vs Truth Seekers	23.126	0.000	Careful vs Private	23.867	0.000
Neutral vs Truth Seekers	14.211	0.000	Neutral vs Private	7.527	0.006
Truth Seekers vs Carefree	135.592	0.000	Private vs Carefree	92.759	0.000
Careful vs Neutral	115.895	0.000	Careful vs Neutral	102.789	0.000
Careful vs Carefree	366.194	0.000	Careful vs Carefree	291.021	0.000
Neutral vs Carefree	78.462	0.000	Neutral vs Carefree	62.852	0.000

5. Digital communication between young people and parents: a tale of two studies

Jodi Dworkin and Xiaoran Sun

5.1 INTRODUCTION

Research on parenting has generally focused on what parents do to children. However, scholarship has increasingly highlighted that parenting is bidirectional, meaning that both parents and children influence each other's behaviours. In fact, much of parenting is a reaction to a child's behaviour (Kuczynski, 2003). From this perspective, we might hypothesize that adolescents and young adults, having grown up using technology unlike their parents, initiate technology use, and parents then respond to that use by also using those same digital tools for communication. In essence, young people are assumed to be agents or active initiators of their technology use with family. This is important to consider as technology has been recognized as playing an increasingly important role in shaping adolescents' agency (Sun, 2024) – a psychological construct that is central to their development towards autonomy and independence from parents (e.g., Beyers et al., 2003).

Indeed, the literature has recognized the "technology broker" role of young people in the family (Correa, 2014; Dworkin, 2018), given young people are the "digital natives" compared with their parents who are more likely to be considered "digital immigrants" (Nelissen & Van den Bulck, 2018). However, there is little research exploring adolescent agency in their digital communication with parents – do adolescents initiate and parents simply respond, or are parents also initiators of communication? In a study of Chilean families, Correa (2014) found some evidence that young people were influencing parents' use of technology, with cautions that those findings were not universal across all families. It is essential to explore whether it is the parent or child who tends to initiate the technology use, and why they choose to initiate – is it an intentional choice to communicate about something specific that may be difficult to talk about in person, or is it a choice based on what they think their communication partner prefers? Do parents choose to communicate via technology because they think that their teen only wants to communicate via technology?

While it is clear that both parents and adolescents frequently use technology to communicate, data also suggest that technology communication does not replace in-person communication but rather supports and enhances it, ultimately contributing to a stronger parent–child relationship. More digital communication has been found to be associated with reports of greater parent–child closeness (Jensen et al., 2021a). Still, when given a choice, young people prefer to communicate in person rather than via technology with parents (Connell & Dworkin, 2012).

There is still a lot to learn about parent–adolescent digital communication and how it impacts family relationships. Both children and parents are navigating how to use digital technology within their relationships. Young people are both active agents in using technology and recipients of their parents' technology use. In this chapter, we start to untangle some of this complexity based on two studies, each using a unique dataset in the US.

More specifically, the two studies collaboratively address agency among young people as it is manifested in their digital communication with parents. In the first study, we use parent–adolescent dyadic survey data to investigate how frequently adolescents and parents choose to use different technologies to communicate with each other, how they feel when they use technologies, and why they choose to use or avoid technology. Next, using intensive objective smartphone observations of young adults' communication with their parents, we explore who sends or initiates more communications, and what patterns of parent–young adult communication look like over time. The self-reported subjective data from Study 1 and the smartphone-based objective observations from Study 2 complement each other methodologically and together illuminate young people's agency and parent–child relationship dynamics in the digital context.

5.2 STUDY 1: FREQUENCY AND PERSPECTIVES ON PARENT–ADOLESCENT DIGITAL COMMUNICATION

In this first study, we consider how frequently parents and adolescents choose to use digital communication, how it makes them feel when they do choose technology to communicate, and why they choose to use technology. We also consider whether the frequency of digital communication is associated with parent–child relationship quality.

5.2.1 Sample and Method

A national US sample of parents and their adolescent children aged 13–18 (404 dyads) was recruited using Qualtrics online recruitment services in late 2019 as part of a larger study. Specifically, Qualtrics aggregated over 20 market research panels and randomly selected potential respondents who were parents of adolescents to receive an invitation to participate in this study. Potential participants were emailed an invitation to participate in a research survey, including the expected length of the survey and the incentives available to them (e.g., cash, airline miles, gift cards). The content of the survey was not revealed until the consent process to reduce self-selection bias.

Parents who met inclusion criteria (i.e., were a parent of a 13–18-year-old) and agreed to participate were required to provide consent before completing the survey. They then completed a 20-minute online survey in Qualtrics. Parents responded to questions regarding their youngest child aged 13–18. Parents were then invited to provide consent for the adolescent (aged 13–18) they reported on to complete a child survey; if they consented, they provided an email address for their child. An email with a link to a child survey was sent to the adolescent. Parent participants were compensated by Qualtrics up to $3.00 in the way they are typically compensated. Adolescents were compensated with a $10 Amazon gift card. This online approach allowed us to recruit a diverse sample of families from across the US who were actively using technology.

5.2.2 Choosing to Use or Avoid Technology: How Does it Make You Feel?

When asked how frequently (1 = never, 7 = several times a day) they communicate with each other via a series of different technology tools, parents and adolescents agree on the frequencies of in-person communication (M = 6.66 vs. 6.70, respectively), phone calls (5.63 vs. 5.55, respectively), and texting (6.03 vs. 5.91, respectively). Mothers ($M = 6.74$) reported communicating with their adolescent in person significantly more frequently than fathers ($M = 6.53$; $t = -2.05$, $p < 0.05$). In contrast, parents and adolescents disagree on frequencies of email (3.54 vs. 3.21, $p < 0.001$, respectively), posting on social media (4.36 vs. 4.06, $p < 0.01$, respectively) and communicating via online video (3.49 vs. 3.28, $p < 0.05$, respectively), with parents reporting significantly more frequent communication than adolescents. It is perhaps not surprising that there is more discrepancy between their reports for email and social media posts, which can be less dyadic and interactive; parents may email or post to interact with their adolescent, but adolescents may not always respond when initiated by a parent and may initiate those communication methods less frequently.

These are also the tools used less frequently, generally once a month or less, potentially subject to greater recall bias than more frequently used tools.

When asked to consider how they feel when communicating via technology, both parents and adolescents reported that communicating via video and text made them feel happy and connected, and more than one-third of parents reported that communicating via text helped them feel informed, which is particularly important for parents who are working to keep track of their adolescents' daily lives. One-third of parents and adolescents also reported that video made them feel excited (see Table 5.1). More mothers than fathers reported that text-based communication made them feel happy ($x^2 = 5.14, p < 0.05$); more fathers than mothers reported that video-based communication made them feel valued ($x^2 = 4.98$, $p < 0.05$). Neither parents nor adolescents associated texting or video communication with feeling worried or overwhelmed (see Table 5.1). Overall, parent–adolescent digital communication was associated with positive feelings (i.e., happy, connected, valued, excited), with different tools serving different purposes for adolescents, mothers, and fathers.

About half of both parents and adolescents reported frequently or very frequently (see Tables 5.2a and b) choosing technology to avoid having to talk in person, to talk about difficult things, to say something they would rather not say directly, or because it feels easier. Fewer than one-third of parents reported choosing technology because they feel that the only way to talk to their child is by using technology. Young people can be the drivers of which technologies are used to communicate with parents. Parents may not be comfortable with the technology, but if they feel it is the only way their child wants to talk, they may choose to use it; however, this is true in the minority of families. Additionally, only about one-third of parents and adolescents reported

Table 5.1 *How using video and text to talk with family makes parents and adolescents feel*

-	Video (% yes): parent	Video (% yes): adolescent	Text (% yes): parent	Text (% yes): adolescent
Happy	59.3	59.8	56.5	58.3
Connected	57.5	61.7	63.7	64.7
Valued	25.7	25.7	22.5	26.7
Excited	32.6	35.1	22.7	29.1
Worried	1.7	2.2	2.0	2.2
Overwhelmed	4.0	5.4	3.2	3.7

Table 5.2a *How frequently parents and adolescents choose to use technology*

Response	Parent	Adolescent
To talk about things I find difficult to talk about face-to-face	45.6	57.1
To tell something to someone that I'd rather not say directly	40.9	51.5
To avoid missing out on something	52.7	60.1
Because I find it easier to communicate online than in real life	49.3	54.1
Because the only way to talk to my child is by using technology	29.3	–

avoiding technology because they feel kids are too distracted by technology and that it prevents them from connecting with others (see Tables 5.2a and b).

Three-quarters of parents and adolescents prefer in-person communication for receiving or offering advice or recommendations and receiving or offering emotional support. Even for making plans, which can often be easily done via text, more than half of parents and adolescents preferred to do that in person (see Table 5.3). Mothers, as compared with fathers, reported preferring in-person communication to receive or offer emotional support ($x^2 = 17.95, p < 0.05$). Mothers preferred in-person communication to inform others about changes in their life ($x^2 = 18.30, p < 0.05$); fathers (though still low; 9.2 per cent vs. 6.8 per cent) reported a preference for texting for this purpose.

These findings are consistent with prior research, which suggests that despite high rates of technology use, young people still choose to communicate in person when it is about more important or serious matters (Connell & Dworkin, 2012). Most technology communication is neutral in sentiment (Sun,

Table 5.2b *How frequently parents and adolescents choose to avoid technology*

Response	Parent	Adolescent
Because it distracts me from really connecting with others	30.4	22.6
Because my child is too distracted by technology	31.8	–
Because kids my age are too distracted by technology	–	31.8

Table 5.3 Preference for communicating in person, via phone or text

When communicating with [parent/child], which method would you choose to:	Parent (%): in person	Parent (%): phone	Parent (%): text	Child (%): in person	Child (%): phone	Child (%): text
Receive or offer advice or recommendations	74.9	2.7	4.2	78.4	4.2	7.2
Receive or offer emotional support	76.7	5.7	2.2	79.9	5.5	4.2
Make plans	57.3	12.4	20.3	62.5	9.7	18.0

Dworkin, et al., 2023), further supporting the idea that technology may be well suited to more mundane topics or check-ins, but more important topics are still discussed in person.

Finally, we conducted regression analyses to examine whether, after controlling for age and gender, the frequency of parent or adolescent reports of digital communication with their adolescent or parent, respectively, would be associated with the quality of the parent–adolescent relationship. There were no significant findings. The pathways through which parent–child digital communication is associated with the quality of the relationship are likely complex; between- and within-family differences might depend on individual and family characteristics, as well as their patterns of digital communication. We need a more nuanced understanding of technology use beyond frequency to begin to understand who is choosing which tools for which conversations and then explore what happens during those conversations.

In sum, this study includes data that reflects why parents and adolescents choose to communicate using technology and how technology makes them feel when they do choose to use it to communicate. The data reveal that parents and adolescents report similar choices when using technology, reasons for choosing technology, and how they feel when using technology. However, these findings also suggest that in order to deepen our understanding of parent–child digital communication, the field needs more innovative and complex data collection techniques that move beyond self-report surveys, which are subject to recall biases and lack details. Innovative research methods such as Screenomics move us towards capturing actual communication patterns over time, as well as identifying who initiates communication without relying on self-reports of frequency.

5.3 STUDY 2: DIRECTIONS OF COMMUNICATION BETWEEN YOUNG ADULTS AND THEIR PARENTS VIA SMARTPHONES – A SCREENOMICS APPROACH

In this section, we showcase the directions of digital communication between parents and young adults using objective and high-intensity observational data collected through the Screenomics approach (Reeves et al., 2020; Sun, Ram, et al., 2023). The direction of the communication – whether sent/initiated by the young adult or the parent – can be a reflection of young adults' agency in their communication with parents.

5.3.1 Sample and Method

Screenomics is an IRB-approved, privacy-protected, and secure approach that captures, collects, and stores screenshots from the smartphones of participants who consented to participate in the study, every 5 seconds over weeks and months (Reeves et al., 2020). For the current analysis, we focused on a sample of 10 young adults who participated in a Screenomics study in 2017. The study was focused on collecting data from adults for up to about a month in three metropolitan areas in the US (New York City, Chicago, and Los Angeles), among which we selected a gender-balanced (50 per cent female, 50 per cent male) and racially diverse (30 per cent White, 20 per cent African American, 20 per cent Hispanic, 20 per cent Asian, 10 per cent Multiracial) subsample of 10 young adults aged between 18 to 23 years for the current analysis.

The selected young adults participated in the study for 6 to 30 days (median = 26.5 days) and contributed a total from 16,331 to 106,862 screenshots. We processed and coded these screenshots using a mixed-method approach combining (1) Optical Character Recognition (OCR), a computer vision module for text extraction from each screenshot; (2) keyword search for communication instances with parents based on the OCR text; and (3) human labelling for the time, channel (e.g., texting, calling, video call, email), parent identity (mother or father), and direction (i.e., from or to parent) of each filtered communication instance. For a detailed description of this method, please see Sun, Ram, et al. (2023).

The mixed-method coding identified screenshots containing communication instances with parents, ranging from 20 to 1,780 screenshots per young adult. Based on these screenshots, we identified a range of four to 320 communication instances across the young adults. Among a total of 1,403 communication instances, 1,032 were messages (including 646 SMS messages and 386 messages through online messaging apps such as WhatsApp), 359 were calls (including 349 phone calls, seven online voice calls, and three online video calls), and 12 were through other communication channels (including two voicemails, two Facebook interactions, six emails, and two location sharing instances).

5.3.2 Direction of Communication Instances between Young Adults and their Parents at the Person Level

Table 5.4 shows the number of observed days and the frequency and percentage of each young adult's communication instances with their parents by parent (mother/father) and direction (to/from parent). Overall, across the whole participation period and all the communication instances, for all young adults but one (YA02), it appeared that the young adult received smartphone communication more frequently from their mother than the communication that they sent to their mother. This aligns with the existing literature that mothers tend to be the primary caregivers and kin-keepers (Thompson & Walker, 1989) – even in the digital context (Sun & McMillan, 2018). In other words, even though young people tend to be the digital brokers and use smartphones for communication more frequently, in family communication, mothers still seem to dominate more than the young adults.

In contrast to the patterns regarding communication with mothers, communication with fathers via smartphones shows a broader range of patterns across young adults. For three (YA01, YA07, YA09) out of the ten young adults, we could not detect any communication on their smartphones with their father. Among those for whom we detected communication with fathers, two young adults (YA02, YA06) sent significantly more communication instances to their fathers than received from them; one (YA03) received many more communications from their father than they sent to him; and four (YA04, YA05, YA08, YA10) received almost equal numbers (+/−2) of instances from their fathers compared with what they sent. These varied patterns indicate that between young adults and their fathers, there is not a clear dominant pattern where either of them leans more towards communicating via smartphones.

Furthermore, at the person level across the entire period of participation, we also looked into the direction of communication between young adults and their mothers and fathers specifically for messages and calls – the two most common channels of family smartphone communication detected in our data (see Table 5.5). Starting with communication with mothers, the dominating direction could differ across different channels. For example, although for both YA03 and YA08, they received more messages from their mothers than they sent to their mothers, they called their mothers more often than their mothers called them. Further, although YA06 did not have any messaging communication with their mother, calls were frequent, with them receiving

Table 5.4 *Days of smartphone communication observations, total number of communication instances with parents, and frequency of communication instances to/from mother/father among the young adult participants*

Participant ID	Observed days	Total number of communication instances directly with parents	To mother	From mother	To father	From father
YA01	9	4	1 (25.0%)	3 (75.0%)	0	0
YA02	6	97	25 (25.8%)	21 (21.6%)	31 (31.9%)	21 (21.6%)
YA03	30	181	62 (34.2%)	82 (45.3%)	9 (5.0%)	28 (15.5%)
YA04	29	275	9 (3.3%)	13 (4.7%)	126 (45.8%)	127 (46.2%)
YA05	24	319	77 (24.1%)	221 (69.3%)	11 (3.4%)	10 (3.1%)
YA06	30	59	23 (40.0%)	31 (52.5%)	5 (8.5%)	0
YA07	30	206	101 (49.0%)	105 (51.0%)	0	0
YA08	29	40	19 (47.5%)	19 (47.5%)	2 (5.0%)	0
YA09	22	19	7 (36.8%)	12 (63.2%)	0	0
YA10	8	38	10 (26.3%)	17 (44.7%)	5 (13.2%)	6 (15.8%)

Note: N (%) in the four right-hand columns denotes total communication instances to/from each parent.

Table 5.5 *Frequency of messages and calls to/from mother/father for each young adult participant*

Participant ID	Total number of messages directly with parents	To mother	From mother	To father	From father	Total number of calls directly with parents	To mother	From mother	To father	From father
YA01	0	–	–	–	–	4	1 (25.0%)	3 (75.0%)	0	0
YA02	96	23 (23.9%)	21 (21.9%)	31 (32.3%)	21 (21.9%)	1	1 (100%)	0	0	0
YA03	156	50 (32.0%)	78 (50.0%)	5 (3.2%)	23 (14.7%)	25	12 (48.0%)	4 (16.0%)	4 (16.0%)	5 (20.0%)
YA04	165	4 (2.4%)	7 (4.2%)	90 (54.5%)	64 (38.8%)	110	5 (4.5%)	6 (5.5%)	36 (32.7%)	63 (57.3%)
YA05	278	75 (27.0%)	182 (65.5%)	11 (3.9%)	10 (3.6%)	39	2 (5.1%)	37 (94.9%)	0	0
YA06	1	0	0	1 (100%)	0	58	23 (39.7%)	31 (53.4%)	4 (6.9%)	0
YA07	314	100 (31.8%)	214 (68.2%)	0	0	2	1 (50.0%)	1 (50.0%)	0	0
YA08	22	8 (36.4%)	14 (63.6%)	0	0	18	11 (61.1%)	5 (27.8%)	2 (11.1%)	0
YA09	15	6 (40.0%)	9 (60.0%)	0	0	2	1 (50.0%)	1 (50.0%)	0	0
YA10	25	7 (28.0%)	10 (40.0%)	3 (12.0%)	5 (20.0%)	5	0	4 (80.0%)	0	1 (20.0%)

Note: N (%) in columns 3, 4, 5, 6, 8, 9, 10, and 11 denotes total communication instances to/from each parent.

more calls from their mother than they made to their mother. For some young adults, however, the pattern was consistent across messages and calls – such as for YA05 and YA10 – their mothers both sent more messages to them and called them more often than they sent messages and calls to their mothers.

Turning to the messages and calls with their father, we also observed different patterns across various communication channels. For example, YA02 sent more messages to their father than they received from him, but there were no calls observed either to or from their father. In contrast, YA03 received more messages from their father than they sent, but the calls to and from their father were quite close. Furthermore, YA04 sent more messages to their father than they received, but their father called them much more often than they called him.

Overall, for communication with both mother and father, the examples demonstrate that the pattern of who dominated more in the communication – whether the young adult or the mother/father – differs depending on which communication channel is being used. However, despite the general stereotype that young people may prefer messaging/texting over calling, in our data we could not find evidence consistently showing that in family communication, young adults would message more or call less than their parents. In other words, there is idiosyncrasy in their family smartphone-based communication regarding which communication channel to use and with whom they communicate with.

5.3.3 Directions of Communication Instances between Young Adults and their Parents at the Daily Level

Beyond the patterns summarized at the person level across the entire participation, the fine granularity of the data (i.e., collected every 5 seconds) also allowed us to observe and analyse smartphone behaviours and communication patterns in more detail. In this section, we zoom into each young adult's communication instances to/from their mother and father at the daily level, as depicted in Figure 5.1. First, these plots show the day-to-day variability in communication frequency with parents. Additionally, they also show how the direction of communication with each parent may vary across different days. For example, although overall YA03 received more communication instances from their mother than those sent to the mother – which is also the pattern across many days of their participation (e.g., Days 1, 3, 10, 11, 26, and 31) – there were still days when YA03 sent more instances to their mother, such as Days 2 and 21. The plot for YA04 shows day-to-day variability in the direction of communication between the young adult and their father. There were days when the instances sent to their father were more than those received from their father, such as Days 13, 16, 17, 21, 23, and 28; however, there were also

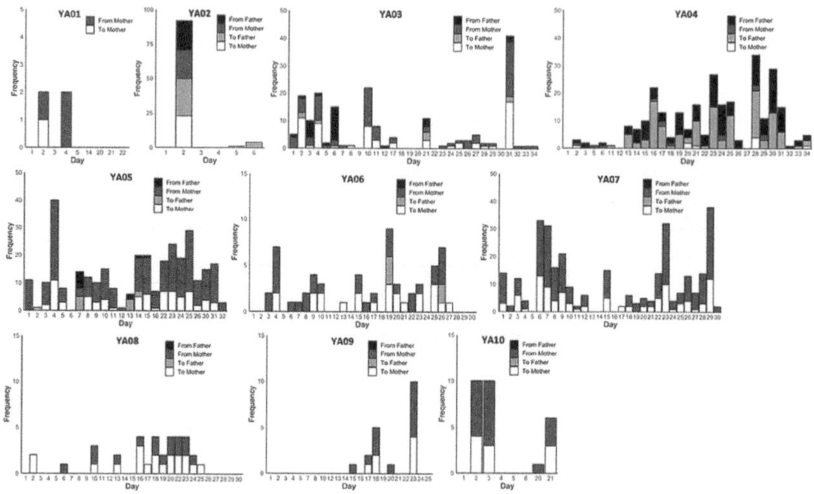

Figure 5.1 *Frequency of communication instances from/to mother and father for each young adult on each participation day*

days when they received more from their father than those sent to their father, such as Days 14, 15, 19, 26, and 30. These variabilities may manifest changes in their relationship dynamics that are worth further examination.

Despite the day-to-day variability in communication patterns regarding the direction of the communication instances, interestingly, there also seems to be consistency in this pattern for some young adults. For example, it is clear that for both YA05, YA07, and YA10, across almost all participation days, they received more communication instances from their mothers than those sent to their mothers. This overtime consistency may represent routinized family communication, but may also demonstrate parental control and overinvolvement as well as consistent power struggles between the young adult and the parent – especially the mother.

5.3.4 Qualitative Analysis of Missed Calls

Among the calls captured in our data between young adults and parents, some calls were missed calls, either initiated by the young adult and missed by the parent or initiated by a parent and missed by the young adult. Descriptively, young adults in our sample were more likely to miss calls from their parents (n = 45) than to have the calls they initiated by them missed by their parents (n = 15).

Because we were able to observe in situ what was happening on the young adult's screen, we also conducted qualitative investigations into the cases where the young adults missed calls from their parents and what screen activities they engaged in before and after the missed calls. Specifically, among the 45 calls missed by the young adults, we observed the smartphone's lock screen before and after 35 calls, suggesting that in these cases the young adults were not engaging with their screen and most likely missed the parent's call due to inattention to their smartphone. However, for the other 10 calls, the young adult was clearly using their smartphone while the parent's call was coming in, engaging in activities including mobile games, watching YouTube videos, and texting with friends. This shows that even though the majority of the missed calls seemed accidental, there were still cases where young adults seemed to intentionally not accept the call from their parents, indicating their intention to reject the communication.

Beyond the screenshots right before and after the missed calls within the same phone use session, we also looked into the screenshots for the phone use session after the session containing the missed calls. Here, we define each session as a consecutive period of phone use from screen on to screen off. Among the 35 calls that were most likely missed due to inattention, we observed another call from the same parent following six of the 35 missed calls, and the young adults either calling or texting back to the parent following six of these calls. Furthermore, following nine of these calls, we found that the young adult engaged in smartphone communication behaviours with social partners (e.g., texting with friends and romantic partners), other than the parent. For the rest of the calls ($n = 14$), young adults engaged in different smartphone use behaviours such as checking social media and utility apps.

In summary, in call-based communication, we found that young adults were more likely to miss calls from their parents than their parents were with them. Even though most of these missed calls appeared to be unintentional, the young adults only followed up with their parents for a small proportion of the missed calls. This overall pattern suggests that young adults seem to not be proactive or initiating in their smartphone communication with their parents, and sometimes are even relatively passive and rejecting.

5.4 DISCUSSION

In this chapter, we combine two unique datasets to begin to depict a more comprehensive picture of parent–child digital communication, considering young people as both agents of technology use and as being influenced by the use of others, in this case, parents. Specifically, an important contribution of these studies is the use of multiple methods and the focus on the intricacies of unique parent–child relationships. For instance, these data reveal that we must not assume mother–child and father–child digital communication are the same. Study 1 suggests that parents' and adolescents' choice of when to use technology to communicate may not be that different. In fact, both parents and adolescents report feeling happy and connected when communicating via text and video-based digital communication. Study 2 findings remind us that interactions within a relationship can change daily or even from hour to hour. While sometimes young people choose to use technology, other times they may choose to avoid technology, and other times their interactions with technology are more passive – they may miss a notification and consequently do not respond to a call or text. Youth do not seem to be particularly agentic in initiating digital communication with their parents. This is critical for family practitioners; to best support family relationships, they must understand how, why, and when parents and children are using technology to communicate. Prior studies have found that, in general, adolescents text and call their parents relatively infrequently, but on those days when there is more digital communication, in-person interactions are more positive (Jensen et al., 2021a). Further, more frequent digital communications have not been found to detract from time spent interacting in person (Jensen et al., 2021b).

There seem to be many assumptions about adolescent and young adult technology use, some of which are correct and some of which have not yet been confirmed. For instance, we have assumed that young people are agents in the family system and likely the leaders and initiators in technology use in the family, for instance, by introducing new technologies to the family and preferring to communicate via technology. These studies suggest that the story is much more complicated and nuanced, and depends on the relationship, reason for communicating, and channel of digital communication.

5.4.1 Practical Implications

Understanding how, why, and when young people and their parents communicate via digital technology is critical for providing family support. Digital communication is a frequent means of communication for families, allowing them to stay connected when they are not able to talk in person or when they are uncomfortable talking in person. There are assumptions about how parents and young people wish to communicate and why, but little data to support those assumptions. These two studies contribute to our understanding of how and why parents and children communicate digitally and the outcomes of that communication, which is essential to supporting healthy and positive family communication.

5.4.2 Avenues for Future Research

Together, these two studies provide important directions for future research. First, in Study 1, findings are cross-sectional and primarily descriptive. Thus, changes in communication over time or differences in emotional responses to in-person and digital communication were not among the focus areas of this research. Therefore, it is essential that research on technology use recognize the historical time and place during which the data were collected. COVID-19 has changed expectations for technology use and, for many families, altered how they use technology and rely on it for school, work, and entertainment. Additionally, qualitative data would allow a deeper exploration of both parents' and adolescents' perspectives on their communication and choices about communication.

Second, in Study 2, the number of observed days was variable across participants – from 6 days to 30 days. It is likely that among young adults who participated in the study for fewer days, the pattern of their communication with parents during those days was less representative of their common pattern. Additionally, the larger study was mainly focused on adults' behaviours on the screen. Therefore, future studies could also gather detailed information about their parents or their family structures. The communication patterns could differ significantly across young adults depending on the parents' marital/partnered status and the family's living arrangements.

Third, given the relatively small sample size for both studies, we were unable to statistically test individual differences, including those across adolescents and young adults and families from different cultural backgrounds. This highlights the importance of collecting and examining larger samples in future research.

Fourth, the parents included in these two studies generally did not grow up with technology in the same way their children did. As younger generations who did grow up with technology as an essential part of their daily lives start to become parents, there will likely be a generational shift in both how individuals use technology and also how they choose to parent around technology. We will need to document these shifts over time to see how this changes technology use within the family and also how it changes parenting around technology use (Dworkin et al., 2018).

REFERENCES

Beyers, W., Goossens, L., Vansant, I., & Moors, E. (2003). A structural model of autonomy in middle and late adolescence: Connectedness, separation, detachment, and agency. *Journal of Youth and Adolescence*, *32*, 351–365. https://doi.org/10.1023/A:1024922031510.

Connell, J., & Dworkin, J. (2012). College students' information and communications technology (ICT) use with parents. *Journal of the Association of Higher Education Parent/Family Program Professionals*, *2*(2), 2–17.

Correa, T. (2014). Bottom-up technology transmission within families: Exploring how youths influence their parents' digital media use with dyadic data. *Journal of Communication*, *64*, 103–124.

Dworkin, J. (2018). Adolescents as the family technology innovators. In J. Van Hook, S. M. McHale, & V. King (Eds), *National symposium on family issues: Vol. 9. Families and technology* (pp. 23–36). Springer.

Dworkin, J., Rudi, J., & Hessel, H. (2018). The state of family research and social media. *Journal of Family Theory & Review*, *10*(4), 796–813. https://doi.org/10.1111/jftr.12295.

Jensen, M., George, M. J., Russell, M. A., Lippold, M. A., & Odgers, C. L. (2021a). Daily parent adolescent digital exchanges. *Research on Child and Adolescent Psychopathology*, *49*, 1125–1138. https://doi.org/10.1007/s10802-020-00765-x.

Jensen, M., George, M. J., Russell, M. A., Lippold, M. A., & Odgers, C. L. (2021b). Does adolescent digital technology use detract from the parent-adolescent relationship? *Journal of Research on Adolescence*, *31*(2), 469–481. https://doi.org/10.1111/jora.12618.

Kuczynski, L. (2003). *Handbook of dynamics in parent–child relations*. Sage.

Nelissen, S., & Van den Bulck, J. (2018). When digital natives instruct digital immigrants: Active guidance of parental media use by children and conflict in the family. *Information, Communication & Society*, *21*(3), 375–387.

Reeves, B., Robinson, T., & Ram, N. (2020). Time for the human screenome project. *Nature*, *577*(7790), 314–317. https://doi.org/10.1038/d41586-020-00032-5.

Sun, X. (2024). The role of digital technology in adolescent autonomy development: A four-factor model and a global perspective. In Y. R. Xia, M. R. T. de Guzman, R. Esteinou, & C. S. Hollist (Eds), *Global perspectives on adolescents and their families. International and cultural psychology* (pp. 21–38). Springer. https://doi.org/10.1007/978-3-031-49230-3_2.

Sun, X., & McMillan, C. (2018). Interplay between families and technology: Future investigations, In J. Van Hook, S. M. McHale, & V. King (Eds), *National symposium on family issues: Vol. 9. Families and technology* (pp. 177–187). Springer. https://doi.org/10.1007/978-3-319-95540-7_10.

Sun, X., Dworkin, J., & LeBouef, S. (2023). Late adolescents' texting experiences with family: Mixed-method analysis for understanding themes and sentiments. *Adolescents*, *3*(3), 581–593. https://doi.org/10.3390/adolescents3030041.

Sun, X., Ram, N., Reeves, B., Cho, M.-J., Fitzgerald, A., & Robinson, T. N. (2023). Connectedness and independence of young adults and parents in the digital world: Observing smartphone interactions at multiple timescales using Screenomics. *Journal of Social and Personal Relationships*, *40*(4), 1126–1150. https://doi.org/10.1177/02654075221104268.

Thompson, L., & Walker, A. J. (1989). Gender in families: Women and men in marriage, work, and parenthood. *Journal of Marriage and the Family*, *51*(4), 845–871.

6. Everyday digital parenting: datafication, mediation and sharenting

Ana Jorge, Francisca Porfírio and Rita Grácio

6.1 INTRODUCTION

The role of media in everyday life and the home has long been the object of reflection and inquiry in different disciplines (Ytre-Arne, 2023; Chambers, 2016). To reflect on such a role today, however, requires us to acknowledge how portable media are not confined to the home and, in turn, how the home is permeated by permanent connectivity, as well as emergent technologies (Kennedy et al., 2020; Leaver, 2017). Exploring the role of digital media in the lives of families with children, in particular, requires us to pay attention not only to how they are inserted in and create routines and rituals but also to how their uses express perceptions and wider views of the world of those families in such a mediatized, datafied, and algorithmic world (Mascheroni & Siibak, 2021; Das, 2024).

This chapter takes a unique approach by situating children and parents within the context of the pervasiveness and intrusiveness of digital media technologies and seeks to understand how families with children engage with these technologies. Through an ethnographic study, based on observations and interviews, it delves into how parents position themselves and negotiate their views in relation to the institutional and social normalization of digital media in everyday life; mediate their children's use of digital technologies; and represent them through social media. Additionally, it explores how children perceive and negotiate their parents' practices. Throughout, it underscores the entangled influence of structural factors affecting families and agency in their interactions with and around digital media. The chapter's sections articulate literature on each of these topics, drawing mostly on Media Studies, and the findings from our study.

6.2 MATERIALS AND METHODS

The study was conducted under the On&Off research project,[1] which looks at the articulations between digital media connection and disconnection in Portugal. Portugal, a Southern European country, is characterized by a "family welfare society", that is, with low levels of welfare protection and where the family is a central support entity (Matias et al., 2021, p. 348).

We analyse how negotiations of media use and non-use take place in families with at least one child up to 12 years old, seeking diversity (as per, for example, Harden et al., 2010) in terms of education and occupation, family forms and living arrangements, geography, and migrant status. Over the Spring of 2023, one of the authors, at times accompanied by another research assistant, conducted ethnography with five families, including a background questionnaire, observation, and interviews. We advertised the study with an e-flyer among our social networks and invited participants who contributed to our diverse sample. Observation took place on two occasions of three hours each, on regular days and during a family meal, with only notes being produced, not audio or video. On a third visit, interviews were conducted with parents (as couples/pairs, except for the single parent in family Barroso) and with children eight or older; as elicitation, interviews used a photo of the participants' favourite device(s) which was requested at the end of the second visit. Our study involved 20 participants – of whom, nine adults and seven children were interviewed (Table 6.1). A total of around 9 hours of interviews were transcribed verbatim, and pictures were blurred for anonymity; all the data was then coded in qualitative analysis software MaxQDA. The codebook for analysis was deductive (based on concrete theoretical and epistemological positions – DeCuir-Gunby et al., 2011) and inductive (a researcher developed inductive codes from the data itself); and the coding was discussed regularly among the team. The participants were compensated for their efforts through a voucher. While this strategy facilitated access to families' everyday lives, it mainly attracted lower- to middle-class families.[2] Participation in the research was anonymous and confidential as per a consent form signed by parents and adapted for children; we chose pseudonyms (popular Portuguese surnames for ABCDE), as suggested by Gurrieri and Drenten (2019) to reference each of the five participant families, and individual pseudonyms were chosen by the participants or the authors.

6.3 DATAFIED AND DIGITAL PARENTING

Digital media technologies are an inescapable part of the world in which parents are raising their children (Barassi, 2017; Leaver, 2015, 2017; Tiidenberg

Table 6.1 Sample characteristics

Family	Adults (mother/ father, age)	Adults' education and occupation	Family form	Area of residency	Migrant status	Children (gender, age)
Almeida	Sandra (mother, 41) Hugo (father, 44)	Education assistant Bus driver and firefighter	Married	Rural (central mainland Portugal)	Both Portuguese	Clara (F, 9) Simão (M, 16)
Barroso	Ana (mother, 33)	Technical Assistant (public entity)	Divorced, mother as sole carer	Rural (central mainland Portugal)	Portuguese	Matilde (F, 7) Margarida (F, 11)
Carvalho	Graça (mother, 35) João (father, 38)	Photographer and content creator IT technician	Common-law marriage	Urban (Great Lisbon)	Both Portuguese	Diana (F, 1)
Dias	Dulce (mother, 47) Júlio (father, 41)	Technical Assistant (private entity) Cook	Common-law marriage after divorce	Urban (Great Lisbon)	Immigrant father	Child of Dulce and Júlio: Ângelo (M, 5) Dulce's children: Lisa (F, 11), Daniel (M, 16) Júlio's child: Guilherme (M, 12)
Esteves	Lina (mother, 39) João (father, 49)	Salesperson Audiovisual technician Both unemployed	Common-law marriage	Urban	Both Portuguese	Vasco (M, 4) André (M, 8)

& Baym, 2017), or even before conceiving them (Lupton & Pedersen, 2016). Underpinning the datafication of children's and adults' lives is the socially accepted assumption that real-time recording and predictive analytics can improve people's lives and help them manage various domains of their lives, including parenting (Barassi, 2017). Even before children have their personal devices, they are connected to their parents in several ways to tend to their "homework, individual needs, school-related activities, as well as social events involving their children" (Lim, 2016, p. 23), in what has been labelled by Lim as "transcendent parenting".

Google Classroom and kindergarten/school apps have become indispensable communication tools between educational institutions and parents, as shown by Cino et al. (2021) and Hooper et al. (2022). In our sample, parents of young children perceive good value in these apps as they can help communicate with these institutions, which can, in turn, be strategic for parents to deal with children's behaviour (for instance, if a child has not slept much at school, this can justify why they are irritable at home). Parents are also made to feel that they would miss out on this information about their children if they opted out; however, no critical reflections emerged related to the datafication that these apps produce.

João, father to one-year-old Diana, tells us that the kindergarten uses an app where "it's very easy for us to see the history of poops, how much she's eaten, or what time she's had her nap and when something needs to be taken to the nursery when there's an event". He adds that "the kindergarten proposed it and it's very useful". Also, Júlio finds the leisure centre app helpful in knowing more about when the child arrived at the facilities, how they behaved on a particular day, etc.

Not only the age of the children but also the parenthood experience is relevant to account for the different uses of apps. Lina Esteves, mother of two children four years apart, recounts she had a health app from the private clinic with foetal monitoring and reminders, but quickly adds: "I think there are things that happen with the first child; then with the second one, it's different".

Fertility or pregnancy apps (Lupton & Pedersen, 2016) were mostly referred to by mothers, but their use might also be negotiated among both parents. Graça Carvalho, Diana's mother, thinks that fertility apps are close to useless in planning for a pregnancy, while her partner João sees that they create additional anxiety when the couple is trying to conceive. This example also shows that, while João sees positive benefits in the kindergarten app, he thinks that fertility apps might be more detrimental than helpful to the family.

Instant Messaging apps have become naturalized for communication between childcare, educational, or recreational institutions and parents, among parents and their peers, as well as with the extended family. In the Italian context, Cino and colleagues (2021) found that parenting WhatsApp groups tended to increase from preschool to elementary school, peaking in secondary school and decreasing after that. These instant messaging groups show that parenting is increasingly a digital endeavour (Mascheroni et al., 2018), and that matters pertaining to child-rearing are not bound to physical and immediate contexts (Lim, 2016; 2019).

In our study, fathers and mothers particularly mentioned WhatsApp groups. This type of group has been found to accentuate social norms and expectations related to the constant use of digital media (Bayer et al., 2016), sometimes accompanied by pressure for people to be available (Geber et al., 2023). These social norms and expectations can be related to the philosophy of "intensive parenting", a highly demanding, child-centred style that holds parents responsible for most of their children's outcomes in life (Faircloth, 2021). Our results reveal that mothers appear to be more attentive to and intervening in these chat groups. Sandra (Almeida) says she pays special attention to groups connected to children's football activities, which have become part of the routine, over the "family" groups, which sit more in the background. Dulce (Dias) expresses willingness to be constantly online so she can be updated "in real time" about "groups for the children, for the leisure centre, for the school", regarding her three children and her partner's son. Although Dulce mentions that this constant availability and monitoring of multiple instant messaging groups is done "without pressure", it seems to add to the burden of mothers' labour and responsibility.

6.4 POSITIONS

We look at how parents position themselves towards media in the household in negotiation with wider social discourses regarding parenthood as well as broader social discourses about media themselves, drawing on Haddon (2013). Previous literature has revealed how parents play a significant role in shaping children's practices with media (Mascheroni & Haddon, 2015), and that they often see digital media as positive and challenging at the same time (Kumpulainen & Gillen, 2017; Chaudron et al., 2017). We draw on Sukk and Siibak's (2021) typology of pre-teens in relation to parental intimate surveillance through tracking apps (studied by Leaver, 2017), which comprises three categories: the tech-trusting, the cautious, and the careful authoritarian parent. We expanded this typology to make sense of the wider digital media use – of both parents and children.

Most of the parents (six) in our sample were "cautious". Sandra (Almeida), Graça and João (Carvalho), Júlio (Dias), João and Lina (Esteves) believe that the media can give a false sense of security and pose risks to children's privacy, which is why they think they should keep an eye on their children. In their view, this practice does not limit children's privacy. João (Esteves) says he has used tracking apps but prefers to talk with his son: "If it's something serious, I'll explain to him why I think it's not good content for him". In contrast, Hugo (Almeida) was the only "tech-trusting" parent who trusts the media and believes that its use does not limit their children's privacy.

We found two "careful authoritarian" mothers – Dulce (Dias) and Ana (Barroso) – who perceive that the media pose significant privacy risks and find it difficult to trust children in their use of media technologies. According to Sukk and Siibak (2021), the profile of careful authoritarian parents is associated with a feeling of insecurity on the part of parents about the media and problems in trusting children during use, whereby adults believe they have the right to control their children. Therefore, they resort to surveillance through supervision apps and parental control features (Ghosh et al., 2018), and/or direct control mechanisms. Dulce illustrates the use of direct control mechanisms as she follows her children on Instagram and says, "I often go to see what they're up to"; she and her partner also keep the children's social media passwords. Similarly, Ana checks her eldest daughter's phone often and has the Instagram password to access it; she uses the supervision app Family Link and Nintendo's parental control app. This is not done without dilemmas; Dulce concedes that "I know it might be a bit of an invasion of their privacy, but it's because of their age". In Ana's case, there is some pondering: "it's very difficult to balance, isn't it? Her [Margarida's] privacy with her safety, because sometimes I have to […], sometimes I have to go a little over her privacy to protect her. Maybe it's not right, but it's the way I protect her." As a single mother, Ana's decisions are more constrained by her lack of time to engage with both her children's media use (Russo Lemor, 2006). In recalling how she dealt with getting a phone for Margarida, she says she got support or information from someone outside the family: "When I wanted to buy the phone, I looked into various parental control apps. I heard about Family Link from a friend, and we installed it." She explains that she has been more permissive over time: "Then I opened doors and gave her permissions that she didn't have initially because Family Link can be as restrictive as possible, and restrictive to the point where she can't even install an application without asking my permission". In explaining it in this way, Ana seems to indicate that she could be more restrictive.

From Sukk and Siibak's (2021) typology applied to our study, we stress that mothers and fathers do not always align and have the same profile (Brito et al., 2017). Interestingly, in couples with different individual positions, namely Dulce and Júlio, and Sandra and Hugo Almeida, the mothers' position tends to prevail in relation to children's media use, in either case to more constraining dispositions than the fathers'. In our study and across the socio-demographic differences, mothers appear to be the primary carers for children, including in overseeing digital parenting. While it can be a marker of Southern European culture, this centrality of mothers in child-rearing has also been found elsewhere, for example, in Switzerland (Balleys, 2022).

Therefore, negotiations between mothers and fathers occur in a gendered context, that is, in a context where both men and women experience the "second shift" (Hochschild & Machung, 1989), but women are expected to take on most of this responsibility in social reproduction. Like Balleys' study (2022), in our cases, when interviewed, children in three families mentioned the mother as being responsible for regulating screen time on a daily basis. In the Esteves family, the mother, Lina, does not allow "technology at mealtimes" and tells us that "André, on his school nights, knows that after he's done his homework, he can use his technology until dinner time".

6.5 NEGOTIATIONS

With regard to how the children react to the limits imposed by their parents, especially their mothers, most children in our sample were "compliant", considering their parents to have the right to total control over their actions in the media and that their safety depends on such control, which they accept. This was the case for Clara (Almeida), Margarida (Barroso), and André (Esteves). Margarida, for instance, seems to have internalized her mother's careful authoritarian position. She told us that her mother has access to her computer and that, while playing on her smartphone, her mother would watch over her, even when she is cooking or watching a series. To Margarida, this is "not a problem". The compliance of younger children can be attributed to the fact that children do not perceive parental control, or that parents see them doing many offline activities, which was especially the case in families with outdoor spaces in their houses and in rural areas. Screen time is a focus of discussion for parents and young children, as Clara, the nine-year-old girl from the Almeida family, tells us:

> My mother is already looking into installing it [referring to the parental supervision app]. There's a friend of mine, Leandro, his dad has this thing to check his stuff on his cell phone. I've already seen him spend 4 hours on TikTok, 4 hours on TikTok in a row.

While younger children tended to be more compliant, older children in our sample adopted an autonomous position, which parents confirmed. Three children – Simão (Almeida), 16; sister Lisa and brother Guilherme (Dias), 11 and 12, respectively – had an "autonomous" profile, characterized by self-awareness, believing that their parents should not constantly control them without their knowledge, as the child is responsible for their own safety. Lisa told us that she would "rather [the mother or the stepfather] weren't there" looking at what she is doing on the phone. Lastly, a "privacy-sensitive" profile was found in Daniel (Dias), a child with some reservations about the media, who preferred to "be alone in my room" using it and take responsibility for their own safety and well-being.

This autonomy is also relevant in how each child says they would react if they did not have access to the internet for one day. Clara, 9, indicated that she would "cope", though she would miss it. André, 8, recalled that he had already experienced a similar situation when he was in a hotel, but that he felt happy because there were lots of activities and he met new people. This example shows how leisure opportunities create opportunities for disconnection, as has also been identified in Jorge et al. (2023). It was young teenagers who showed slightly more negative feelings towards this scenario: Lisa (Dias) said that she would feel angry if she did not have internet for a week, and her brother Guilherme said he would only consider not using the internet for a day if he got paid for it. Margarida said she could go a couple of days without using the internet, but not a whole week. Older participants, Simão Almeida and Daniel Dias, both 16, seemed to react better to the idea of technology being absent, though temporarily. The autonomy that children gain in their teenage years seemed to be postponed for girls and for children of the sole mother, in the Barroso family. Margarida tells us that her mother "got upset because I spent a lot of time in the office without leaving my office and as it's downstairs, it heats up quickly. And she got angry with me and then I couldn't play [on the PC] for two days [...] I shouldn't have played for so long." On the contrary, it appeared to be anticipated by children with older siblings. In our observations, we saw five-year-old Ângelo asking for help from his older siblings and his father to type what he wanted to search on YouTube when the audio search was not available on the smart TV. Júlio, his father, is aware and concerned that older siblings might also cause younger Ângelo to have contact with age-inappropriate content: he told the boy "not to watch something you're not supposed to" on YouTube, explaining to the researcher that "he must have watched it with his siblings".

Siblings' interactions might also be an occasion for learning and reciprocity (Siibak & Nevski, 2019), which we detected when seven-year-old Matilde sat with her older sister Margarida to see what she was doing online. As a younger sibling, Matilde inherited her sister's computer when she upgraded. We also found instances where older siblings are seen as tech savvy. For example, Sandra told us that it was Simão who blocked someone on Instagram who the younger sister had added unintentionally. In some cases, if children of close ages are available, interactions between siblings might be suspended. In our observation in the Esteves' family house, the father, João, remarked that it was exceptional that André was playing with his younger brother and a girl neighbour close to his age in the shared backyard. If researchers had not been there, he said André would have been "plugged into the machine".

Previous studies (Balleys, 2022; Chaudron et al., 2017) have noted that children circulate in different environments and agreements of norms. In our study, this situation was especially evident in the case of Dias' children, a common-law marriage after divorce. Guilherme tells us:

> I'm my mum's only child. She just says, "put it away". And if I don't put it away, my mum says, "what did I tell you?" and I say "[you said] put it away" and my mum gets this look on her face. And I get really scared, so I just switch it off and stop [...] but with Aunt Dulce [stepmother] I just use a bit more.

Lina (Esteves) provides further examples that demonstrate that, when children also engage with other relatives, rules might change: "if it's with me, it's one thing... If it's with their father, it's another. If grandma is here, they know they can get around another way."

Rarely did we see or hear about clashes between parents and children regarding media use during our data collection, which we can attribute to a desirability effect. However, children and parents often referred to children's attempts to use devices a little longer than the parents would have liked. As Lina says: "Although I don't think they're very difficult children in that sense, because they're very accepting of our... but they're children, they're always trying to find the limit".

6.6 TO SHARE OR NOT TO SHARE

Despite revealing concerns about children's privacy, parents frequently share information about their children on digital media (Choi & Lewallen, 2018), which has been dubbed the "privacy paradox" (Ní Bhroin et al., 2022). In fact, within social media culture, children are easily represented in photos, videos, memes, and other forms of content by their parents and others. Over the last ten years, the literature on sharenting – sharing parenting – has focused mainly on the practices performed by mothers (Fox & Hoy, 2019), who have the traditional role of the "storytellers" of families (Humphreys, 2018). However, fathers are increasingly turning to social media as a space of visibility to share their family representations through their narratives of fatherhood (Campana et al., 2020). Tartari et al. (2023) argue that sharing is a gender issue because mothers are more willing to share, but, according to Mascheroni et al. (2023, p. 113), "the gender dimension of sharing is not a given".

In general, in our sample, mothers share more often than fathers. Sharenting practices appear normalized, trivialized even, and unproblematic for the participant parents. They mainly refer to motivations related to pride, memory, and *doing* family, either of times spent together or to feel like a family by perpetuating those memories, as can be seen in the following quotes:

> I think sometimes it's also a question of pride, for example, when there's something different about our children that we like everyone to see. (Sandra Almeida)

> If we go anywhere, to different places, we take it down and put it up [on social media]. (Hugo Almeida)

> When I share, it's not for the views, it's not for the likes. I share it because I think it's something I like to share. The way I think about it is, I'm sharing with my community. (Ana Barroso)

> The advantage is the continuation of the family, seeing our children grow up, because you're not with them [...] a way for all of us to be more united, when we're not. (Lina Esteves)

When contemplating the risks associated with sharenting, Ana (Barroso) says that the sheer volume of children's photos online, posted by parents as well as institutions, actually protects her children in the photos she shares of them on social media. However, parents are aware of the replicability and scalability of online photos. With this in mind, some mothers indicated that they practice what Autenrieth (2018) has labelled "anti-sharenting", for example, using digital manipulation to post about the children without sharing their location, identifiable elements, or private parts. This practice reduces the focus on the image of the children themselves and emphasizes their spatial context. Ana does this when she shares a picture of her seven-year-old girl topless at the beach, "I'll use some emoji to cover her nipples". Parents also expressed general concern for the consequences over children's safety and, overall, seemed to have incorporated safety precautions circulated by media and educational programmes. João Esteves, for instance, tells us that he shares content of his children "with respect for the rules, the photographs, not identifying things, I have no reference to schools. And [...] we don't share other people's."

Again, we found that tensions may arise with different positions of the carers. For instance, in the Carvalho family, the father indicates that he does not share photos of his daughter, which can be seen as a form of non-sharing, a type of practice portrayed by Damkjaer (2018) as being characteristic of fathers with a higher level of education. He states, "If it were up to me, Diana wouldn't exist on the internet. However, Diana is part of our lives, part of Graça's life. And Graça wants to share part of her life. It's done with common sense and agreement." In turn, the mother is a small-size creator of digital content, especially on Instagram, and likes to include her baby in her content, always with her partner's permission. She explains that "we don't expose Diana; fortunately, we have the money to pay for the things she needs". While she accepts gifts and sees that people like to see the child on social media, she associates children's exposure with the monetary gains of influencers. What Graça does resent is the unsolicited advice from followers. She tells us that just by briefly mentioning "I'm frustrated by yet another ear infection", "then I get 1,000 messages saying, 'oh, you should do this, you should do that' and I've already done it…".

Children in our study did not demonstrate that they perceive sharenting as something negative. Only André had conditional approval, as he said that he would not mind "unless it was something embarrassing". These results are similar to those of Ouvrein and Verswijvel's (2019) study with 12 to 14-year-olds, which showed that adolescents generally approve of sharing and trust their parents, but do not exclude the possibility that sharing can create embarrassing situations. However, this might be due to a lack of awareness and communication. Lisa, 11, told us that her mother did not share images of her on social media, which was different from what the mother, Dulce, told us in her interview. Again, the older participants, Simão Almeida and Daniel Dias, were the ones to express more indifference towards what parents shared about them online, which means that privacy and independence can have very different meanings depending on age, as Ouvrein and Verswijvel (2019) pointed out.

6.7 DISCUSSION

Our study foregrounds how parents and children deal with the intensive digitalization pushed by institutional and social environments into which they are inserted. In the context of parenting, digital media are inexorable – in fact, ever-expanding – elements in everyday family life, coordinating child-rearing responsibilities and communicating with extended family, friends, and others related to children. In particular, our findings call attention to the co-surveillance enabled by data technologies, be it between parents and children, or among parents and other institutions. While this form of surveillance might be, in instances, a social protection for children, it also raises critical questions about increasing control over them and limiting their autonomy and agency. Parents negotiate their uncertainties – and disagreements – about digital media, both in how they mediate their children's use and how they represent them online, by incorporating and modulating circulating discourses about risks vis-à-vis their motivations, views about technology, and parenting styles.

Parents held different positions in relation to digital media, affected by education, economic conditions, parenthood experience, and, not least, time to attend to children's different needs (Russo Lemor, 2006), which were evident in our mostly low- and middle-class sample. In our study, it was especially mothers who mediated children's media use and maintained the labour of transcendent parenting (Lim, 2019), as well as sharenting (Humphreys, 2018). Furthermore, there were significant geographical differences among the families; rural areas and privately owned outdoor spaces, or access to safe public spaces, increased children's possibilities to interact with others and play, and take digital media as an alternative. Negotiations among parents and children about the use of digital media are affected by their age or developmental stage.

6.7.1 Practical Implications

While families are involved in broader processes of datafication and digitalization, including circulating discourses about them, and parents play a pivotal role in structuring children's engagement with digital media, we argue that those processes do not determine families' relationships to them. Besides their influence by – and on – other agents, notably peers and the school, children hold forms of agency in how they resist and recreate the space for digital media in their lives, for example, in how they interpret different contexts with different forms of mediation. Nevertheless, to empower children and families, we insist on public policies for city planning that allow for public and accessible spaces for children, and welfare policies to offer them more diverse leisure options. There is an urgency to reach new agreements and norms – by schools and communities – about digitalization regarding family life that are considerate of their needs and hold the best interests of children at the centre.

6.7.2 Avenues for Future Research

It would be fruitful if future research from other cultural settings could reveal more about the configurations of media and leisure in everyday families' lives, particularly exploring how different cultures situate the agency of parents and children. This can also be explored through other research avenues, such as longitudinal research, returning to families 5 and 10 years later to explore how evolved media conditions, but especially their grown children, have impacted their perceptions.

NOTES

1. DOI:10.54499/2022.01282.PTDC.
2. Although we sought these groups, some of the middle- to upper-class families who responded to our flyer ad eventually decided not to take part, seeing the voucher as little compensation for their commitment to the study.

REFERENCES

Autenrieth, U. (2018). Family photography in a networked age: Anti-sharenting as a reaction to risk assessment and behaviour adaption. In G. Mascheroni, C. Ponte, & A. Jorge (Eds), *Digital Parenting: The Challenges for Families in the Digital Age* (pp. 219–231). Nordicom.

Balleys, C. (2022). Familial digital mediation as a gendered issue between parents. *Media, Culture & Society*, *44*(8), 1559–1575. https://doi.org/10.1177/01634437221119020.

Barassi, V. (2017). BabyVeillance? Expecting parents, online surveillance and the cultural specificity of pregnancy apps. *Social Media+ Society*, *3*(2), 2056305117707188. https://doi.org/10.1177/2056305117707188.

Bayer, J. B., Ellison, N. B., Schoenebeck, S. Y., & Falk, E. B. (2016). Sharing the small moments: Ephemeral social interaction on Snapchat. *Information, Communication & Society*, *19*(7), 956–977. https://doi.org/10.1080/1369118X.2015.1084349.

Brito, R., Francisco, R., Dias, P., & Chaudron, S. (2017). Family dynamics in digital homes: The role played by parental mediation in young children's digital practices around 14 European countries. *Contemporary Family Therapy*, *39*(4), 271–280. https://doi.org/10.1007/s10591-017-9431-0.

Campana, M., Bossche, A. V. den, & Miller, B. (2020). #dadtribe: Performing sharenting labour to commercialise involved fatherhood. *Journal of Macromarketing*, *40*(4), 475–491. https://doi.org/10.1177/0276146720933334.

Chambers, D. (2016). *Changing Media, Homes and Households: Cultures, Technologies and Meanings*. Routledge.

Chaudron, S., Di Gioia, R., & Gemo, M. (2017). *Young Children (0–8) and Digital Technology: A Qualitative Study across Europe*. EUR 29070 EN, Publications Office of the European Union, Luxembourg. doi:10.2760/294383.

Choi, G. Y., & Lewallen, J. (2018). "Say Instagram, kids!": Examining sharenting and children's digital representations on Instagram. *Howard Journal of Communications*, *29*(2), 144–164.

Cino, D., Gigli, A., & Demozzi, S. (2021). "That's the only place where you can get this information today!". An exploratory study on Parenting WhatsApp Groups with a sample of Italian parents. *Studi sulla Formazione/Open Journal of Education*, *24*(1), 75–96.

Damkjaer, M. S. (2018). Sharenting=good parenting?: Four parental approaches to sharenting on Facebook. In G. Mascheroni, C. Ponte, & A. Jorge (Eds), *Digital Parenting: The Challenges for Families in the Digital Age* (pp. 209–218). Nordicom.

Das, R. (2024). *Parents Talking Algorithms: Stories of Parenthood in an Algorithm Age*. Bristol University Press.

DeCuir-Gunby, J. T., Marshall, P. L., & McCulloch, A. W. (2011). Developing and using a codebook for the analysis of interview data: An example from a professional development research project. *Field Methods*, *23*(2), 136–155. https://doi.org/10.1177/1525822X10388468.

Faircloth, C. (2021). When equal partners become unequal parents: Couple relationships and intensive parenting culture. *Families, Relationships and Societies*, *10*(2), 231–248. https://doi.org/10.1332/204674319X15761552010506.

Fox, A. K., & Hoy, M. G. (2019). Smart devices, smart decisions? Implications of parents' sharenting for children's online privacy: An investigation of mothers.

Journal of Public Policy & Marketing, 38(4), 414–432. https://doi.org/10.1177/0743915619858290.

Geber, S., Nguyen, M. H., & Büchi, M. (2023). Conflicting norms: How norms of disconnection and availability correlate with digital media use across generations. *Social Science Computer Review*. https://doi.org/10.1177/08944393231215457.

Ghosh, A. K., Badillo-Urquiola, K., Guha, S., LaViola Jr, J. J., & Wisniewski, P. J. (2018, April). Safety vs. surveillance: What children have to say about mobile apps for parental control. In *Proceedings of the 2018 CHI Conference on Human Factors in Computing Systems* (pp. 1–14).

Gurrieri, L., & Drenten, J. (2019). Visual storytelling and vulnerable health care consumers: Normalising practices and social support through Instagram. *Journal of Services Marketing, 33*(6), 702–720. https://doi.org/10.1108/JSM-09-2018-0262.

Haddon, L. (2013). Mobile media and children. *Mobile Media & Communication, 1*(1), 89–95. https://doi.org/10.1177/2050157912459504.

Harden, J., Backett-Milburn, K., Hill, M., & MacLean, A. (2010). Oh, what a tangled web we weave: Experiences of doing "multiple perspectives" research in families. *International Journal of Social Research Methodology, 13*(5), 441–452. https://doi.org/10.1080/13645571003650979.

Hochschild, A., & Machung, A. (1989). *The Second Shift: Working Families and the Revolution at Home*. Penguin Books.

Hooper, L., Livingstone, S., and Pothong, K. (2022) *Problems with Data Governance in UK Schools: The Cases of Google Classroom and ClassDojo*. Digital Futures Commission, 5Rights Foundation.

Humphreys, L. (2018). *The Qualified Self: Social Media and the Accounting of Everyday Life*. The MIT Press.

Jorge, A., Agai, M., Dias, P., & Martinho, L. C.-V. (2023). Growing out of overconnection: The process of dis/connecting among Norwegian and Portuguese teenagers. *New Media & Society*, 14614448231159308. https://doi.org/10.1177/14614448231159308.

Kennedy, J., Arnold, M., Gibbs, M., Nansen, B., & Wilken, R. (2020). *Digital Domesticity: Media, Materiality, and Home Life*. Oxford University Press.

Kumpulainen, K., & Gillen, J. (2017). Young children's digital literacy practices in the home: A review of the literature. *Cost Action ISI1410 DigiLitEY*. https://digiliteymethodscorner.files.wordpress.com/2018/10/wg1-kumpulainen_gillen_lit-review-2017.pdf.

Leaver, T. (2015). Born digital? Presence, privacy, and intimate surveillance. In J. Hartley & W. Qu (Eds), *Re-orientation: Translingual Transcultural Transmedia: Studies in Narrative, Language, Identity, and Knowledge* (pp. 149–160). Fudan University Press.

Leaver, T. (2017). Intimate surveillance: Normalizing parental monitoring and mediation of infants online. *Social Media + Society, 3*(2), 2056305117707192. https://doi.org/10.1177/2056305117707192.

Lim, S. S. (2016). Through the tablet glass: Mobile media, cloud computing and transcendent parenting. *Journal of Children & Media, 10*(1), 21–29. https://doi.org/10.1080/17482798.2015.1121896.

Lim, S. S. (2019). *Transcendent Parenting: Raising Children in the Digital Age*. Oxford University Press.

Lupton, D., & Pedersen, S. (2016). An Australian survey of women's use of pregnancy and parenting apps. *Women and Birth, 29*(4), 368–375. https://doi.org/10.1016/j.wombi.2016.01.008.

Mascheroni, G., & Haddon, L. (2015). Children, risks, and the mobile internet. In Z. Yan (Ed.), *Encyclopedia of Mobile Phone Behavior* (pp. 1409–1418). IGI Global.
Mascheroni, G., & Siibak, A. (2021). *Datafied Childhoods: Data Practices and Imaginaries in Children's Lives*. Peter Lang.
Mascheroni, G., Cino, D., Amadori, G., & Zaffaroni, L. G. (2023). (Non-)Sharing as a form of maternal care? The ambiguous meanings of sharenting for mothers of 0-to-8-year-old children. *Italian Sociological Review*, *13*(1), 111–130. https://doi.org/10.13136/isr.v13i1.635.
Mascheroni, G., Ponte, C., & Jorge, A. (Eds). (2018). *Digital Parenting: The Challenges for Families in the Digital Age*. Nordicom.
Matias, M., Monteiro, I., & Coimbra, S. (2021). A brief overview of research in Portugal. In I. Albert, M. Emirhafizovic, C.-N. Shpigelman, & U. Trummer (Eds), *Families and Family Values in Society and Culture* (pp. 345–357). Information Age Publishing.
Ní Bhroin, N., Dinh, T., Thiel, K., Lampert, C., Staksrud, E., & Ólafsson, K. (2022). The privacy paradox by proxy: Considering predictors of sharenting. *Media and Communication*, *10*(1), 371–383. https://doi.org/10.17645/mac.v10i1.4858.
Ouvrein, G., & Verswijvel, K. (2019). Sharenting: Parental adoration or public humiliation? A focus group study on adolescents' experiences with sharenting against the background of their own impression management. *Children and Youth Services Review*, *99*, 319–327. https://doi.org/10.1016/j.childyouth.2019.02.011.
Russo Lemor, A. M. (2006). Making a "home". The domestication of information and communication technologies in single parents' households. In T. Berker, M. Hartmann, Y. Punie, & K. Ward (Eds), *Domestication of Media and Technologies* (pp. 165–184). Open University Press.
Siibak, A., & Nevski, E. (2019). Older siblings as mediators of infants' and toddlers' (digital) media use. In O. Erstad, R. Flewitt, B. Kümmerling-Meibauer, & Í. Susana Pereira (Eds), *The Routledge Handbook of Digital Literacies in Early Childhood* (pp. 123–133). Routledge.
Sukk, M., & Siibak, A. (2021). Caring dataveillance and the construction of "good parenting": Estonian parents' and pre-teens' reflections on the use of tracking technologies. *Communications*, *46*(3), 446–467. https://doi.org/10.1515/commun-2021-0045.
Tartari, M., Lavorgna, A., & Ugwudike, P. (2023). Share with care: Negotiating children's health and safety in sharenting practices. *Media, Culture & Society*, *45*(7), 1453–1470. https://doi.org/10.1177/01634437231182002.
Tiidenberg, K., & Baym, N. K. (2017). Learn it, buy it, work it: Intensive pregnancy on Instagram. *Social Media+ Society*, *3*(1), 2056305116685108. https://doi.org/10.1177/2056305116685108.
Ytre-Arne, B. (2023). *Media Use in Digital: Everyday Life*. Emerald.

PART II

Young people as digital consumers and financial actors

7. ICT engagement and other factors associated with adolescents' financial literacy: evidence from PISA

Gintautas Silinskas, Kati Laine and Arto K. Ahonen

7.1 INTRODUCTION

Financial competencies are consistently recognized as being among the most essential 21st-century skills (Kunina-Habenicht & Goldhammer, 2020; Moreno-Herrero et al., 2018). These competencies are important because most people have to make daily financial decisions concerning their spending and saving and differentiate among a wide range of products, services, and providers (Arellano et al., 2014; Moreno-Herrero et al., 2018). For young people, gaining financial independence is one of the key components during their transition to adulthood (Luhr, 2018), and they seem to be aware of this (Ali et al., 2014). Young people also represent an attractive market for financial service providers; owing to their limited experience, young people are often at a greater risk of making poor financial decisions (Arellano et al., 2014). Given this, adolescents should be exposed to education in *financial literacy* – the ability to understand financial information and effectively manage individual finances (Arellano et al., 2018). Moreover, financial markets are increasingly accessible to consumers, due to major technological progress, and people need digital skills for making financial decisions online. Thus, *information and communication technology (ICT) literacy* – the ability to deal with ICT to effectively participate in society (Lennon et al., 2003) – is closely related to financial literacy and thus represents a crucial aspect of children's learning (Kunina-Habenicht & Goldhammer, 2020). Altogether, young people's ability to make financial decisions, especially in digital environments, and thus act towards their own welfare can be seen as a manifestation of their financial agency (DeJaeghere et al., 2016).

The Organisation for Economic Co-operation and Development (OECD) added financial literacy as an option to be assessed in the Programme for International Student Assessment (PISA) in 2012. To date, the optional financial literacy assessment has been available four times in this triennial study. In PISA 2018, 20 countries participated in the computer-based assessment of financial literacy. Besides the cognitive assessment, the PISA study includes non-cognitive background questionnaires for students: the Financial Literacy Questionnaire, the ICT Familiarity Questionnaire, and the Student Questionnaire.[1] We analysed data from the European OECD countries that used these three questionnaires. The countries in this set were assumed to be geopolitically similar (e.g., location and membership in the European Union and OECD). Our analysis included eight countries: three scoring statistically significantly above the OECD financial literacy average (Estonia, Finland, and Poland), one not differing from the OECD average (Latvia), and four countries scoring statistically significantly below the OECD average (Lithuania, Italy, Spain, and the Slovak Republic). The latter feature assured representation across different levels of financial literacy proficiency scores.

In this chapter, we present empirical evidence on the factors predicting financial literacy scores among 15-year-olds in the PISA 2018 assessment. The factors represent five broad categories (see Figure 7.1): (1) ICT competence and autonomy (perceived competence in ICT and perceived autonomy related to ICT use); (2) financial confidence and autonomy (perceived confidence in using digital financial services and perceived autonomy in handling money); (3) home and school learning environments (learning financial concepts at

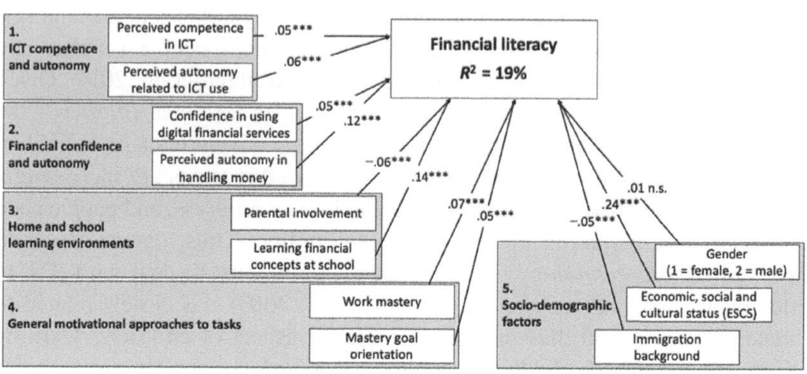

Note: $n = 30{,}019$; n.s. = not significant, $^{***}p < 0.001$.

Figure 7.1 *Predicting financial literacy for the pooled sample of eight European countries*

school and parental involvement in matters of financial literacy); (4) general motivational approaches to tasks (work mastery and mastery goal orientation); and (5) socio-demographic factors (student gender; economic, social and cultural status [ESCS]; and immigration background).

7.2 THEORETICAL FRAMEWORK

Based on the OECD framework (OECD, 2019), we operationalize our dependent variable of financial literacy as the cognitive factor (also, objective PISA 2018 financial literacy score). Financial literacy has been defined as the knowledge and understanding of financial concepts and risks and the skills to apply such knowledge and understanding to make effective decisions across a range of financial contexts (OECD, 2019). The five broad categories of predictors represent the non-cognitive factors measured by student questionnaires (i.e., subjective evaluations; see OECD, 2019). Their importance and expected associations with the financial literacy scores are presented next.

7.2.1 ICT Competence and Autonomy

ICT engagement is a critical factor in ICT literacy and is instrumental in facilitating learning and acquiring new knowledge (Zylka et al., 2015). Among the dimensions of ICT engagement, the roles of perceived competence in ICT and perceived autonomy related to ICT use have been emphasized (see also Goldhammer et al., 2017; Kunina-Habenicht & Goldhammer, 2020). First, *perceived competence in ICT* refers to the beliefs about an individual's knowledge and skills about ICT and the ability to use this knowledge and skills in practice (Goldhammer et al., 2017; Kunina-Habenicht & Goldhammer, 2020). Based on a study by Hatlevik et al. (2018), which utilized International Computer and Information Literacy Study (ICILS) data, experience with ICT and being able to independently learn ICT-based activities are important elements for the development of a student's ICT self-efficacy – that is, perceived ICT competence.

Second, *perceived autonomy related to ICT use* refers to an individual's perceptions of control and self-directedness in ICT activities (Goldhammer et al., 2017; Kunina-Habenicht & Goldhammer, 2020). This positive self-concept and self-regulated ICT behaviour increases the chances of a successful future in dealing with ICT, keeping up with the latest technological developments and, thus, applying ICT knowledge and skills in related domains such as financial literacy, especially in the digital domain. According to Hu et al. (2018), students who reported higher interest, autonomy and competence in ICT exhibited significantly higher performance in mathematics, reading and science in PISA 2015. In addition, Areepattamannil and Santos (2019), utilizing PISA

2015 data, found that both perceived competence in ICT usage and perceived autonomy related to ICT usage were positively associated with science-related dispositions (e.g., enjoyment of science). Overall, given previous research and the strong connection between ICT and financial literacy competencies, we expected that perceived competence in ICT (Hypothesis 1a) and perceived autonomy related to ICT use (Hypothesis 1b) would be positively related to financial literacy scores among adolescents.

7.2.2 Financial Confidence and Autonomy

Non-cognitive financial factors, such as perceived confidence in using digital financial services or perceived autonomy in handling money, can contribute to financial literacy scores (OECD, 2019). First, *confidence in using digital financial services* has been described as the motivation to seek information and advice concerning financial matters, to apply this knowledge in practice and to manage emotional and psychological states when making financial decisions about digital financial services (Arellano et al., 2014; OECD, 2019). Overall confidence (not specific to the financial domain) was shown to predict financial literacy in Spanish PISA 2012 data (Arellano et al., 2018). In PISA 2018, owing to the fast rate of digitalization of financial services, the assessment measured two specific types of confidence concerning financial matters: one related to using traditional financial services (e.g., paying bills and understanding bank contracts) and the other related to using digital financial services (e.g., paying with a mobile device and checking bank balance online). To our knowledge, only one earlier study has investigated the difference between the two types of confidence in relation to financial literacy scores (Silinskas et al., 2023). The authors found that Finnish adolescents' confidence in using digital financial services positively and significantly predicted financial literacy scores.

Second, in addition to confidence-related beliefs, actual practices in the financial domain and concrete financial behaviour were suggested to predict financial literacy (Vosylis & Erentaitė, 2020; Xiao et al., 2014). One such practice is *perceived autonomy in handling money*, which has been described as the ability to handle one's own money (OECD, 2020). Based on the OECD report on PISA 2018 (2020), on average, 81 per cent of students agreed or strongly agreed that they could independently decide what to spend their money on. The report also showed that students who were more autonomous in their money matters performed better in the PISA 2018 financial literacy assessment after accounting for their gender, ESCS and immigration background. This measure has not been further investigated in the literature but appears highly relevant at the developmental stage of adolescence, which is typically characterized by increased attempts to expand independence and autonomy from parents (Luhr, 2018; Silinskas & Kikas, 2019; Silinskas et al., 2023). The results of a study on young adults (aged 18–23 years) in the United States indicated that economic (young adult's income, assets, work status and educational attainment) and psychological factors (economic self-efficacy, money management ability and problem-solving ability) were positively associated with their financial independence, whereas family economic factors (parental income, stock holding and financial assistance) were negatively associated with financial independence (Xiao et al., 2014). Taken together, feeling confident in using digital financial services (Hypothesis 2a) as well as practising planning expenses and independently taking care of money matters (Hypothesis 2b) could be expected to be positively related to financial literacy scores among adolescents of OECD countries in Europe.

7.2.3 Home and School Learning Environment Factors

Children learn financial information from their social environments. Thus, discussions with parents about financial matters (Jorgensen & Savla, 2010) and lessons mentioning financial concepts at school are assumed to promote their financial literacy skills. First, across OECD countries, 94 per cent of adolescents, on average, indicated parents as the primary source of information about financial matters (OECD, 2020). Parents may play a role in their children's financial socialization by engaging them in and exposing them to various experiences such as observing and imitating parental behaviour, making financial resources available and creating opportunities to participate in financial practices, and communicating about financial matters (Danes & Yang, 2014; Vosylis & Erentaitė, 2020). PISA 2015 data indicated that parental involvement was positively related to financial literacy scores in nine of the 15 participating countries (Moreno-Herrero et al., 2018). Using PISA 2018 data from Finland, Silinskas et al. (2021) showed that parental involvement may also negatively predict financial literacy scores, perhaps because parental involvement was found to be the least frequent in countries such as Finland and Estonia, where 15-year-olds reported the highest usage of financial services and the highest autonomy in using them (OECD, 2020). Moreover, the developmental stage of adolescence is characterized by an increased need for autonomy and independence from parents (Silinskas & Kikas, 2019; Silinskas et al., 2022). Thus, parental involvement in their children's financial socialization may provide mixed results in relation to financial literacy scores (Hypothesis 3a not specified owing to mixed previous findings).

Second, based on the Opportunity to Learn framework (OTL framework; Elliott & Bartlett, 2016; Schmidt & McKnight, 1995), school context and exposure to curriculum content, in particular, enhance students' academic achievement. The OTL, overall, emphasizes the idea that learning something requires opportunities to learn (Yang Hansen & Strietholt, 2018). It also highlights the importance of ensuring that all students have equal access to high-quality learning opportunities, including resources, support and instruction. In the PISA 2018 Financial Literacy Questionnaire, adolescents were given a list of 18 financial concepts and were asked to report if they had heard of those concepts or learnt about them at school during the previous 12 months. Results from previous PISA assessments seem to support the idea that more extensive promotion of financial literacy at school is related to higher financial literacy scores. For example, the results of PISA 2012 showed that the availability of financial education as such had a significant positive relation with students' financial literacy scores (Cordero et al., 2022). Consequently, we expected to find a positive association in the pooled data across European OECD countries (Hypothesis 3b).

7.2.4 General Motivational Approaches to Tasks

We investigated two types of achievement motivation factors of how adolescents approach learning in general: work mastery and achievement goal orientation (Elliot & McGregor, 2001; Helmreich et al., 1978). *Work mastery* (sometimes labelled persistence) refers to a dispositional motive to work hard to master tasks, improve skills and persist in tasks until they are finished (OECD, 2019; Silinskas & Kikas, 2019). In PISA 2018, work mastery was positively related to adolescents' financial literacy scores, at least in Finland (Silinskas et al., 2021). Given previous studies and the fact that work mastery is widely believed to be a positive aspect of achievement motivation, and is often encouraged, we expected to find positive associations between work mastery and financial literacy scores.

Another achievement motivational construct is *mastery goal orientation*, which can be understood as focusing on mastering a task and having an internal desire to acquire new knowledge and skills (as opposed to the performance goal orientation – learner's concern in demonstrating competence and focusing on receiving positive external feedback for their ability from others; Dweck & Elliot, 1983). In contrast to work mastery, mastery goal orientation has not been examined in relation to financial literacy scores in PISA assessments. However, a positive association has been found in other domains, such as math literacy (Kriegbaum et al., 2015). Consequently, we expected that a learner who persisted in difficult tasks (i.e., was motivated by work mastery) and tended to focus on improvement in the classroom (i.e., pursued mastery goals) would have higher financial literacy scores than the one who did not (Hypotheses 4a and 4b, respectively).

7.2.5 Socio-Demographic Factors

Numerous socio-demographic background factors, such as gender, ESCS and immigration status, have been shown to be associated with financial literacy scores. These socio-demographic factors are central to analysing and interpreting PISA results (OECD, 2020) because they explain a large amount of variation in financial literacy scores and attract considerable attention from scientists, practitioners, policymakers, and society. Thus, they are acknowledged in this chapter as essential contributors when explaining financial literacy but are not our primary focus. To this end, we explain how these factors relate to financial literacy scores but do not raise specific hypotheses concerning them.

First, in PISA 2018, across all 20 participating countries, boys statistically significantly outperformed girls in financial literacy – by two points on average (OECD, 2020). However, surprisingly, boys statistically significantly outperformed girls only in three countries (Italy, Peru and Poland); girls statistically significantly outperformed boys also in three countries (Bulgaria, Georgia and Indonesia), and gender differences were not significant in the remaining 14 countries. Second, disadvantaged children in terms of ESCS scored significantly lower than advantaged children (a score difference of 78 from the OECD average of financial literacy scores) (OECD, 2020). Third, students with immigrant backgrounds scored significantly lower in financial literacy than their non-immigrant peers in a pool of all 20 countries (5-point difference on average; OECD, 2020). In these analyses, the scores were reported after accounting for performance in mathematics and reading. Reports for individual countries revealed that the difference favouring non-immigrants was significant only in three countries (Georgia, Lithuania and the United States).

7.3 DATA AND METHODOLOGY

This study was based on the publicly available data collected in the PISA 2018 assessment (N = 30,019). We used data from eight countries that participated in the financial literacy assessment: Estonia (n = 3,524), Finland (n = 3,342), Italy (n = 6,080), Latvia (n = 2,485), Lithuania (n = 3,010), Poland (n = 3,509), the Slovak Republic (n = 2,561) and Spain (n = 5,508). The countries in this set were assumed to be geopolitically similar, and the participants in these countries conducted the cognitive financial literacy test and answered the non-cognitive questionnaires used in the present study. The first stage of participant selection involved a stratified sample of schools, and the second stage involved a sample of students within schools. PISA-eligible student ages were between 15 years and three months and 16 years and two months: they were born in 2002 or 2003. The most typical modal grades for students in school were nine and ten. Each country collected its own data, and students took the test in the first half of 2018. Detailed descriptions of the variables and items used in the present study are presented in Tables 7A.1a–e.

7.4 RESULTS

In this study, we predicted the financial literacy of 15-year-old students in a pool of eight European OECD countries based on multiple regression analysis. We used IDB Analyzer version 5.0.20 (a macro of SPSS 24.0, provided by the IEA), which was specifically designed for the analysis of PISA data. Altogether, our model can explain 19 per cent of the variation in the financial literacy scores (see Figure 7.1). The results showed that perceived competence and perceived autonomy of ICT use were positively related to financial literacy scores (Hypotheses 1a and 1b supported). Confidence in using digital services and perceived autonomy in handling money were positively related to financial literacy (Hypotheses 2a and 2b supported). Regarding home and school learning environments, learning financial concepts at school was positively (Hypothesis 3b supported), whereas parental involvement was negatively, related to financial literacy (Hypothesis 3a was not previously specified owing to mixed previous findings). As for general motivational approaches to tasks, both work mastery and mastery goal orientation were positively related to financial literacy scores (Hypotheses 4a and 4b supported). For socio-demographic factors, no gender differences were noted, but low ESCS and having an immigrant background were related to low financial literacy scores. In sum, most of the associations were positive. The only exceptions were parental involvement and immigration background, which showed negative relations with financial literacy. For the results concerning each of the eight countries separately, please consult Tables 7A.2a–e.

7.5 DISCUSSION

This chapter investigates the relative importance of factors relating to the financial literacy of adolescents in the PISA 2018 assessment. Adolescents selected for this study represent eight European OECD countries. The results revealed that financial literacy competencies in adolescents can be related to five broad categories of factors: (1) ICT competence and autonomy, (2) financial confidence and autonomy, (3) home and school learning environments, (4) general motivational approaches to tasks, and (5) socio-demographic factors. Overall, the most consistent results suggest that adolescents who were exposed to financial concepts at school and granted autonomy in handling a certain amount of money independently scored highest in financial literacy.

7.5.1 Theoretical Implications

We found that perceived competence and perceived autonomy in using ICT were positively related to financial literacy scores. Financial literacy goes hand in hand with ICT literacy owing to the fast development of digital devices and services in the financial domain. According to ICT engagement theory, individuals who feel more competent in ICT and more autonomous in using ICT have greater chances of developing better ICT skills (Kunina-Habenicht & Goldhammer, 2020; Zylka et al., 2015). Although ICT competence and autonomous usage have been shown to be related to science-related dispositions, to our knowledge, our study is the first to demonstrate their relation to financial literacy scores as well.

Perceived confidence in using digital services was positively related to financial literacy – a result replicating previous findings in one of the countries in focus (Finland; Silinskas et al., 2023). Moreover, in accordance with previous studies (OECD, 2020; Xiao et al., 2014), adolescents' perceived autonomy in handling money was consistently positively related to financial literacy scores not only for the pooled results but also for each participating country. This is a novel finding. Given that the developmental stage of adolescence is characterized by an increased need for autonomy and independence from their parents (Silinskas & Kikas, 2019; Silinskas et al., 2021), we may expect that this also applies in the domain of financial literacy. This result points to the importance of parents' decisions in how much autonomy in handling money they grant to their children.

Regarding home and school learning environments, we measured parental involvement in children's financial education through discussions, and, consistent with the results from Finland (Silinskas et al., 2021), our results showed a negative relation between parental discussion and financial literacy skills. This result may suggest that parents overall cannot effectively transmit their knowledge or may do so in a controlling and autonomy-thwarting way (Silinskas & Kikas, 2019; Silinskas et al., 2021). Taken together, the main results of the present study emphasize that enabling adolescents to make certain financial decisions independently rather than involving them in direct discussion has the potential to increase their financial literacy. Apart from parental involvement, learning financial concepts at school emerged as one of the most critical factors contributing to adolescents' financial literacy scores. This has been suggested by the OTL framework (Elliott & Bartlett, 2016) and empirically confirmed in previous studies (Cordero et al., 2022; Silinskas et al., 2021). We propose that learning through practice and the knowledge provided in school can positively supplement each other when building adolescents' financial literacy.

As for general motivational approaches to tasks, both work mastery and mastery goal orientation were positively related to financial literacy scores.

This is not surprising because personal predisposition, such as work mastery (or persistence in finishing even challenging tasks), is related to better achievement in the financial literacy domain (Arellano et al., 2018; Silinskas et al., 2021). Similarly, a need to master subjects/academic domains (versus studying for external reasons) is related to achievement in other academic domains (e.g., mathematics; Kriegbaum et al., 2015). However, prior to the present study, less was known about how these qualities relate to achievement in the domain of financial literacy.

Finally, regarding more general background characteristics, although no gender differences were noted, low ESCS and having an immigrant background were related to lower financial literacy scores. Interestingly, among all variables, ESCS emerged as clearly the most important predictor for financial literacy scores. Knowing that students with low ESCS and having an immigration background are in a disadvantaged position, more attention should be paid to their education to promote their financial literacy competencies.

7.5.2 Practical Implications

From a practical viewpoint at the policy level, because learning financial concepts at school emerged as among the most important factors affecting financial literacy scores, financial education at school should be encouraged and implemented. Furthermore, along with enhancing financial knowledge, financial education could include face-to-face or virtual simulations of situations wherein students can make their own financial decisions and see the possible consequences of their decisions.

At the family level, to improve adolescents' financial competence, instead of encouraging them to have direct discussions with parents about financial matters, adolescents should be empowered with some independence and autonomy in handling their own money. When adolescents are old enough, they could also be encouraged to take a small-scale summer job to earn their own money and develop their financial agency.

Schools, parents and practitioners dealing with adolescents should also be aware of the importance of building adolescents' competence/confidence and encouraging their autonomous use of ICT and digital financial services. Important themes related to digital financial services may include digital transactions, cryptocurrencies, online banking and investment platforms (see e.g., Zaimovic et al., 2023). An essential part of the autonomous use of ICT and digital financial services is understanding the risks related to them and learning how to safely use these services. Other motivational competencies, such as an approach to mastering (versus avoiding) school subjects and persisting in completing even difficult tasks, should also be encouraged.

In sum, to achieve high-level financial literacy competencies, one should implement formal education in financial literacy as well as provide possibilities for adolescents to make financial decisions in practice. Facilitated by the combination of financial knowledge and practice, especially in digital environments, adolescents can be provided with tools to develop their financial literacy and build their financial agency.

7.5.3 Avenues for Future Research

This chapter has a few limitations associated with the design of PISA that are also characteristic of many other large-scale assessments. First, the data are cross-sectional, making it impossible to determine which factors are antecedents and which are consequences of financial literacy scores. Second, only students completed financial literacy tasks and answered questionnaires, exposing our results to common-source bias (i.e., the same participants provided all the data). Third, except for the PISA financial literacy score itself, all the other measures rely on student reports on questionnaires, making our results susceptible to common method and social desirability biases. In future studies, the associations reported here remain to be investigated by longitudinal and experimental designs. Further, research may benefit from exploring the reports of other respondents (e.g., parents, teachers and classmates) and applying other data collection methods (e.g., observations).

AUTHORS' NOTE

The data that support the findings of this study, the materials and the analysis code are available on request from the authors. This study was not preregistered. We have no conflicts of interest to disclose. The work of Gintautas Silinskas was supported by the Academy of Finland (#296082, #331525, #336148 and #358041). The work of Kati Laine and Arto K. Ahonen was supported by the Academy of Finland (#327243 and #352549).

NOTE

1. For more information, see the PISA 2018 database (www.oecd.org/pisa/data/2018database/).

REFERENCES

Ali, P., McRae, C. H., & Ramsay, I. (2014). Financial literacy and financial decision-making of Australian secondary school students. *Australian Business Law Review, 42*, 228–233. https://ssrn.com/abstract=2490216.

Areepattamannil, S., & Santos, I. M. (2019). Adolescent students' perceived information and communication technology (ICT) competence and autonomy: Examining links to dispositions toward science in 42 countries. *Computers in Human Behavior, 98*, 50–58. https://doi.org/10.1016/j.chb.2019.04.005.

Arellano, A., Cámara, N., & Tuesta, D. (2014). The effect of self-confidence on financial literacy. *BBVA Research WP, 14*(28), 1–25.

Arellano, A., Cámara, N., & Tuesta, D. (2018). Explaining the gender gap in financial literacy: The role of non-cognitive skills. *Economic Notes: Review of Banking, Finance and Monetary Economics, 47*(2–3), 495–518. https://doi.org/10.1111/ecno.12113.

Cordero, J. M., Gil-Izquierdo, M., & Pedraja-Chaparro, F. (2022). Financial education and student financial literacy: A cross-country analysis using PISA 2012 data. *The Social Science Journal, 59*(1), 15–33.

Danes, S. M., & Yang, Y. (2014). Assessment of the use of theories within the *Journal of Financial Counseling and Planning* and the contribution of the family financial socialization conceptual model. *Journal of Financial Counseling and Planning, 25*(1). https://papers.ssrn.com/sol3/papers.cfm?abstract_id=2466560.

DeJaeghere, J. G., McCleary, K. S., & Josić, J. (2016). Conceptualizing youth agency. In J. DeJaeghere, J. Josić, & K. McCleary (Eds.), *Education and youth agency. Advancing responsible adolescent development.* Springer. https://doi.org/10.1007/978-3-319-33344-1_1.

Dweck, C. S., & Elliot, E. S. (1983). Achievement motivation. In E. M. Hetherington (Ed.), *Handbook of child psychology: Vol. 4. Socialization, personality, and social development* (4th ed.). Wiley.

Elliot, A. J., & McGregor, H. A. (2001). A 2 × 2 achievement goal framework. *Journal of Personality and Social Psychology, 80*(3), 501–519. https://doi.org/10.1037/0022-3514.80.3.501.

Elliott, S. N., & Bartlett, B. J. (2016). Opportunity to learn. In Oxford Handbooks Editorial Board (Ed.), *Oxford handbook topics in psychology.* Oxford University Press. https://doi.org/10.1093/oxfordhb/9780199935291.013.70.

Goldhammer, F., Gniewosz, G., & Zylka, J. (2017). ICT engagement in learning environments. In S. Kuger, E. Klieme, N. Jude, & D. Kaplan (Eds.), *Assessing contexts of learning world-wide: Extended context assessment framework and documentation of questionnaire material.* Springer International Publishing.

Hatlevik, O. E., Throndsen, I., Loi, M., & Gudmundsdottir, G. B. (2018). Students' ICT self-efficacy and computer and information literacy: Determinants and relationships. *Computers & Education, 118*, 107–119. http://dx.doi.org/10.1016/J.COMPEDU.2017.11.011.

Helmreich, R. L., Beane, W., Lucker, G. W., & Spence, J. T. (1978). Achievement motivation and scientific attainment. *Personality and Social Psychology Bulletin, 4*(2), 222–226. https://doi.org/10.1177/014616727800400209.

Hu, X., Gong, Y., Lai, C., & Leung, F. K. S. (2018). The relationship between ICT and student literacy in mathematics, reading, and science across 44 countries: A multilevel analysis. *Computers & Education, 125*, 1–13. https://doi.org/10.1016/j.compedu.2018.05.021.

ILO. (2008). International Standard Classification of Occupations 2008 (ISCO-08). International Labour Organization. Available at: https://ilostat.ilo.org/methods/concepts-and-definitions/classification-occupation/. (Accessed 22 April 2025.)

Jorgensen, B. L., & Savla, J. (2010). Financial literacy of young adults: The importance of parental socialization. *Family Relations*, *59*(4), 465–478. https://doi.org/10.1111/j.1741-3729.2010.00616.x.

Kriegbaum, K., Jansen, M., & Spinath, B. (2015). Motivation: A predictor of PISA's mathematical competence beyond intelligence and prior test achievement. *Learning and Individual Differences*, *43*, 140–148. https://doi.org/10.1016/j.lindif.2015.08.026.

Kunina-Habenicht, O., & Goldhammer, F. (2020). ICT engagement: A new construct and its assessment in PISA 2015. *Large-Scale Assessments in Education*, *8*(1), 1–21. https://doi.org/10.1186/s40536-020-00084-z.

Lennon, M., Kirsch, I., Von Davier, M., Wagner, M., & Yamamoto, K. (2003). *Feasibility study for the PISA ICT Literacy Assessment: Report to Network A*. Educational Testing Service.

Luhr, S. (2018). How social class shapes adolescent financial socialization: Understanding differences in the transition to adulthood. *Journal of Family and Economic Issues*, *39*, 457–473. https://doi.org/10.1007/s10834-018-9573-8.

Moreno-Herrero, D., Salas-Velasco, M., & Sánchez-Campillo, J. (2018). Factors that influence the level of financial literacy among young people: The role of parental engagement and students' experiences with money matters. *Children and Youth Services Review*, *95*, 334–351.

OECD. (2019). *PISA 2018 assessment and analytical framework*. PISA, OECD Publishing. https://doi.org/10.1787/b25efab8-en.

OECD. (2020). *PISA 2018 results (volume IV): Are students smart about money?* PISA, OECD Publishing. https://doi.org/10.1787/48ebd1ba-en.

Schmidt, W. H., & McKnight, C. C. (1995). Surveying educational opportunity in mathematics and science: An international perspective. *Educational Evaluation and Policy Analysis*, *17*(3), 337–353. https://doi.org/10.2307/1164511.

Silinskas, G., Ahonen, A. K., & Wilska, T. A. (2021). Financial literacy among Finnish adolescents in PISA 2018: The role of financial learning and dispositional factors. *Large-scale Assessments in Education*, *9*(1), 1–23. https://doi.org/10.1186/s40536-021-00118-0.

Silinskas, G., Ahonen, A. K., & Wilska, T. A. (2023). School and family environments promote adolescents' financial confidence: Indirect paths to financial literacy skills in Finnish PISA 2018. *Journal of Consumer Affairs*, *57*(1), 593–618. https://doi.org/10.1111/joca.12513.

Silinskas, G., & Kikas, E. (2019). Math homework: Parental help and children's academic outcomes. *Contemporary Educational Psychology*, *59*, 101784. https://doi.org/10.1016/j.cedpsych.2019.101784.

Silinskas, G., Viljaranta, J., Poikkeus, A.-M., & Lerkkanen, M.-K. (2022). Maternal homework approach and adolescents' academic skills: The mediating role of task values. *Frontiers in Education*, *7*, 999128. https://doi.org/10.3389/feduc.2022.999128.

UNESCO. (2006). ISCED 1997: International Standard Classification of Education. May 2006 Re-edition. United Nations Educational, Scientific and Cultural Organization. Available at: https://uis.unesco.org/sites/default/files/documents/international-standard-classification-of-education-1997-en_0.pdf. (Accessed 22 April 2025.)

Vosylis, R., & Erentaitė, R. (2020). Linking family financial socialization with its proximal and distal outcomes: Which socialization dimensions matter most for emerging adults' financial identity, financial behaviors, and financial anxiety? *Emerging Adulthood*, *8*(6), 464–475. https://doi.org/10.1177/2167696819856763.

Xiao, J. J., Chatterjee, S., & Kim, J. (2014). Factors associated with financial independence of young adults. *International Journal of Consumer Studies*, 38(4), 394–403. https://doi.org/10.1111/ijcs.12106.

Yang Hansen, K., & Strietholt, R. (2018). Does schooling actually perpetuate educational inequality in mathematics performance? A validity question on the measures of opportunity to learn in PISA. *ZDM Mathematics Education*, *50*(4), 643–658. https://doi.org/10.1007/s11858-018-0935-3.

Zaimovic, A., Torlakovic, A., Arnaut-Berilo, A., Zaimovic, T., Dedovic, L., & Nuhic Meskovic, M. (2023). Mapping financial literacy: A systematic literature review of determinants and recent trends. *Sustainability*, *15*(12), 9358. https://doi.org/10.3390/su15129358.

Zylka, J., Christoph, G., Kroehne, U., Hartig, J., & Goldhammer, F. (2015). Moving beyond cognitive elements of ICT literacy: First evidence on the structure of ICT engagement. *Computers in Human Behavior*, *53*, 149–160. https://doi.org/10.1016/j.chb.2015.07.008.

APPENDIX

Table 7A.1a *Descriptions of variables: ICT competence and autonomy (ICT familiarity questionnaire)*

Variable/Index	No. items	Item stimulus	Response categories	Items
Perceived competence in financial literacy (index formed by OECD)	5	Thinking about your experience with digital media and digital devices, how much do you agree or disagree with the following statements?	4-point scale: 1 = *Strongly disagree* 2 = *Disagree* 3 = *Agree* 4 = *Strongly agree*	• I feel comfortable using the digital devices that I am less familiar with. • If my friends and relatives want to buy new digital devices or applications, I can give them advice. • I feel comfortable using my digital devices at home. • When I come across problems with digital devices, I think I can solve them. • If my friends and relatives have problems with digital devices, I think I can solve them.
Perceived autonomy related to ICT (summed variable constituted for this study)	5	Thinking about your experience with digital media and digital devices, how much do you agree or disagree with the following statements?	4-point scale: 1 = *Strongly disagree* 2 = *Disagree* 3 = *Agree* 4 = *Strongly agree*	• If I need new software, I install it myself. • I read information about digital devices to be independent. • I use digital devices as I want to use them. • If I have a problem with a digital device, I start solving it on my own. • If I need a new application, I choose it myself.

Table 7A.1b Descriptions of variables: financial confidence and autonomy (financial literacy questionnaire)

Variable/Index	No. items	Item stimulus	Response categories	Items
Confidence about financial matters using digital devices (index formed by OECD)	5	When using digital or electronic devices outside of the bank (e.g., at home or in shops), how confident would you feel about doing the following things?	4-point scale: 1 = *Not at all confident* 2 = *Not very confident* *3 = Confident* *4 = Very Confident*	• Transferring money • Keeping track of my balance • Paying with a debit card instead of using cash • Paying with a mobile device (e.g., mobile phone or tablet) instead of using cash • Ensuring the safety of sensitive information when making an electronic payment or using online banking
Perceived autonomy in handling money (summed variable constituted for this study; items 2 and 3 reverse-coded)	4	How much do you agree with the following statements about the way you handle your money?	4-point scale: 1 = *Not at all confident* 2 = *Not very confident* *3 = Confident* *4 = Very Confident*	• I can decide independently what to spend my money on. • I can spend small amounts of my money independently, but for larger amounts, I need to ask my parents or guardians for permission. • I need to ask my parents or guardians for permission before I spend any money on my own. • I am responsible for my own money matters (e.g., for preventing theft).

Table 7A.1c Descriptions of variables: home and school learning environments (student questionnaire)

Variable/Index	No. items	Item stimulus	Response categories	Items
Parental involvement in matters of financial literacy (index formed by OECD)	4	How often do you discuss the following matters with your parents (or guardians or relatives)?	4-point scale: 1 = *Never or hardly ever* 2 = *Once or twice a month* 3 = *Once or twice a week* 4 = *Almost every day*	• Your spending decisions • Your savings decisions • The family budget • Money for things you want to buy
Learning financial concepts at school (sum of 18 items, formulated by OECD)	18	Thinking about school lessons in the last 12 months, have you heard of, or learnt about, the following terms?	3-point scale: 1 = *Never heard of it* 2 = *Heard of it, but I don't recall the meaning* 3 = *Learnt about it, and I know what it means*	e.g., • Interest payment • Compound interest • Exchange rate • Depreciation • Shares/stocks • etc.

Table 7A.1d *Descriptions of variables: general motivational approaches to tasks (student questionnaire)*

Variable/Index	No. items	Item stimulus	Response categories	Items
Work mastery (index formed by OECD)	3	How much do you agree with the following statements about yourself?	4-point scale: 1 = *Strongly disagree* to 4 = *Strongly agree*	• I find satisfaction in working as hard as I can. • Once I start a task, I persist until it is finished. • Part of the enjoyment I get from doing things is when I improve on my previous performances.
Mastery goal orientation (index formed by OECD)	3	Thinking about your goals in school, how true are the following statements for you?	5-point scale: 1 = *Not at all true for me* 2 = *Slightly true for me* 3 = *Moderately true for me* 4 = *Very true for me* 5 = *Extremely true for me*	• My goal is to learn as much as possible. • My goal is to completely master the material presented in my classes. • My goal is to understand the content of my classes as thoroughly as possible.

Table 7A.1e Descriptions of variables: socio-demographic factors (student questionnaire)

Variable/Index	No. items	Item stimulus	Response categories	Items
Gender	1	Are you female or male?	1 = *female* 2 = *male*	-
Economic, social and cultural status, ESCS index (index formed by OECD)	-	Students' responses on parental education (ISCED 1997 coded). Highest occupational status of parents (ISCO and ISEI 2008). Household possessions, including cultural possessions. Home educational resources. Family wealth. ICT resources.	-	-
Immigration background (index formed by OECD)	-	Country-specific items relating to the student's mother's and father's country of birth. Country of birth was (1) the same as the country of assessment or (2) other. The index of immigration background was calculated from these variables.	1 = Native students (those who had at least one parent born in the country of assessment) 2 = Second-generation students (those born in the country of assessment but whose parent(s) were born in another country) 3 = First-generation students (those born outside the country of assessment and whose parents were also born in another country)	In what country were you and your parents born? Country A Country B Country C … etc. Other country

Note: All the mentioned questionnaires are available on the OECD website (https://www.oecd.org/pisa/data/2018database/). All the used items are obtained from the PISA 2018 assessment and analytical framework document (OECD, 2019). The ESCS index is further described only in the PISA 2018 assessment and analytical framework document (OECD, 2019).

Table 7A.2a *Results of multiple regression analyses predicting financial literacy in each OECD European country by ICT competence and autonomy*

Variable	Estonia $R^2 = 0.19$ $n = 3,524$ β	Finland $R^2 = 0.23$ $n = 3,342$ β	Italy $R^2 = 0.11$ $n = 6,080$ β	Latvia $R^2 = 0.20$ $n = 2,485$ β	Lithuania $R^2 = 0.24$ $n = 3,010$ β	Poland $R^2 = 0.21$ $n = 3,509$ β	The Slovak Republic $R^2 = 0.25$ $n = 2,561$ β	Spain $R^2 = 0.13$ $n = 5,508$ β
Perceived competence in ICT	0.04	0.06*	0.01	0.02	0.19***	0.02	0.05	0.02
Perceived autonomy related to ICT	0.05*	0.06*	0.05	0.08**	0.02	0.11***	0.04	0.05

Table 7A.2b Results of multiple regression analyses predicting financial literacy in each OECD European country by financial confidence and autonomy

Variable	Estonia $R^2 = 0.19$ $n = 3,524$ β	Finland $R^2 = 0.23$ $n = 3,342$ β	Italy $R^2 = 0.11$ $n = 6,080$ β	Latvia $R^2 = 0.20$ $n = 2,485$ β	Lithuania $R^2 = 0.24$ $n = 3,010$ β	Poland $R^2 = 0.21$ $n = 3,509$ β	The Slovak Republic $R^2 = 0.25$ $n = 2,561$ β	Spain $R^2 = 0.13$ $n = 5,508$ β
Perceived confidence in financial literacy	0.05*	0.04*	0.03	0.11***	0.06**	0.09***	−0.01	0.05*
Perceived autonomy in handling money	0.16***	0.04*	0.07***	0.18***	0.18***	0.11***	0.17***	0.04*

Table 7A.2c Results of multiple regression analyses predicting financial literacy in each OECD European country by home and school learning environments

Variable	Estonia $R^2 = 0.19$ $n = 3{,}524$ β	Finland $R^2 = 0.23$ $n = 3{,}342$ β	Italy $R^2 = 0.11$ $n = 6{,}080$ β	Latvia $R^2 = 0.20$ $n = 2{,}485$ β	Lithuania $R^2 = 0.24$ $n = 3{,}010$ β	Poland $R^2 = 0.21$ $n = 3{,}509$ β	The Slovak Republic $R^2 = 0.25$ $n = 2{,}561$ β	Spain $R^2 = 0.13$ $n = 5{,}508$ β
Parental involvement	−0.10***	−0.09***	−0.01	−0.07***	−0.03	−0.10***	−0.04	−0.03
Learning financial concepts at school	0.15***	0.25***	0.08***	0.12***	0.10***	0.14***	0.17***	0.16***

Table 7A.2d *Results of multiple regression analyses predicting financial literacy in each OECD European country by general motivational approaches to tasks*

Variable	Estonia $R^2 = 0.19$ $n = 3,524$ β	Finland $R^2 = 0.23$ $n = 3,342$ β	Italy $R^2 = 0.11$ $n = 6,080$ β	Latvia $R^2 = 0.20$ $n = 2,485$ β	Lithuania $R^2 = 0.24$ $n = 3,010$ β	Poland $R^2 = 0.21$ $n = 3,509$ β	The Slovak Republic $R^2 = 0.25$ $n = 2,561$ β	Spain $R^2 = 0.13$ $n = 5,508$ β
Work mastery	0.05*	0.14***	0.11***	0.09***	−0.02	0.07**	0.07**	0.07**
Mastery goal orientation	0.09***	0.01	−0.01	0.02	0.06**	0.05**	0.09***	0.06**

Table 7A.2e *Results of multiple regression analyses predicting financial literacy in each OECD European country by socio-demographic factors*

Variable	Estonia $R^2 = 0.19$ $n = 3{,}524$ β	Finland $R^2 = 0.23$ $n = 3{,}342$ β	Italy $R^2 = 0.11$ $n = 6{,}080$ β	Latvia $R^2 = 0.20$ $n = 2{,}485$ β	Lithuania $R^2 = 0.24$ $n = 3{,}010$ β	Poland $R^2 = 0.21$ $n = 3{,}509$ β	The Slovak Republic $R^2 = 0.25$ $n = 2{,}561$ β	Spain $R^2 = 0.13$ $n = 5{,}508$ β
Gender (1 = female, 2 = male)	0.02	−0.05*	0.09***	−0.05*	−0.01	0.02	0.02	0.00
Economic, social and cultural status	0.21***	0.21***	0.22***	0.23***	0.25***	0.25***	0.29***	0.22***
Immigration background	−0.08***	−0.09***	−0.01	0.02	−0.07**	−0.04	0.09***	−0.05*

Note: *$p<0.05$; **$p<0.01$; ***$p<0.001$.

8. Did financial identity moderate young adults' social media use and financial well-being during COVID-19?

Mette Ranta, Lijun Li and Joyce Serido

8.1 INTRODUCTION

The ubiquity of social media usage in individuals' daily lives cannot be overstated. At the general population level, social media usage is grounded in communication websites that facilitate relationship forming and sharing user-generated content between individuals from different backgrounds, encouraging inquiry and decision-making (Kapoor et al., 2018). Around 4.89 billion people worldwide currently use social media to share content, interact, and exchange information (Statista, 2023). However, current research on understanding "whether, when, or why heightened reliance on social media becomes benefits or drawbacks, especially in times of crisis," is limited (Cho et al., 2023, p. 1). The surge in social media usage during the COVID-19 pandemic has further spurred interest in both the motivations for social media use and its effects on individuals' well-being and social functioning, particularly among young adults (Lee et al., 2022).

Social isolation in the wake of COVID-19 added to young adults' emotional and social distress (Shanahan et al. 2020; Vehkalahti et al., 2021) and thus it is not surprising that their use of social media as a means of social support rose during the pandemic. In addition to increased social media use, the economic impact of the pandemic affected young adults more negatively compared with older adults (Ranta et al., 2020), impeding progress toward financial independence for many young adults (Delbosc & McCarthy, 2021). According to a Bruegel blog post in the early stages of the pandemic, Eurostat data reported a 4.7-point increase in unemployment among younger workers (aged 15–24) (Grzegorczyk & Wolff, 2020). A recent World Bank report (2023) warns of long-lasting impacts in terms of human capital accumulation, earnings, and economic growth among young people, further eroding their financial well-being.

Social networks can serve as sources of both information and support (Cho et al., 2023; Kim et al., 2011) and young adults may rely on social networks for financial information (e.g., how to manage money and where to go for help). If this is so, we ask, does the type of young adults' social network make a difference in terms of their well-being? For instance, some young adults prefer to interact with like-minded individuals (*homophily*), while others are more open to diverse-thinking individuals (*heterophily*) (Kaakinen et al., 2018). Examining this preference in social media use allows us to consider potential sociodemographic differences among young adults in the association between social media use and financial well-being.

Because preferred social media use might also impact the development of individual agency, expose young adults to different types of financial information, and offer varying levels of support, it is possible that social media use may interact with their self-perception within the financial domain, defined as their *financial identity* (Serido et al., 2023). Financial identity is considered one of many domain-specific identities that contribute to an overall sense of self or personal identity (Crocetti , 2018; Negru-Subtirica et al., 2017). Financial identity has been also seen as an internal compass for how young adults navigate financial challenges (Serido et al., 2023).

In this chapter, we investigate young adults' social media use and financial well-being in the context of a rapidly expanding digitalized society and rising pandemic-driven economic uncertainty. Because digital agency and financial agency may each relate to financial well-being, we investigate the potential that young adults' social media use and financial identity interact in the association with financial well-being, to reveal when social media use is, in fact, a benefit or a drawback to financial well-being.

8.2 LITERATURE REVIEW

8.2.1 Homophilous Social Media Use

The concept of *homophily* is used to describe the tendency of individuals to connect with others who are like them, primarily in terms of social identification, background, or attitudes. In principle, people feel more comfortable with people who are like them, and they tend to associate and bond with like-minded people (Kaakinen et al., 2018). Homophily is also defined as forming relationships with similar people (McPherson et al., 2001), that is, with people who share the same background or attitudinal factors (McCroskey et al., 2006). Homophily is also important in friendship formation as individuals tend to be attracted to people who are similar to them (Laakasuo et al., 2020). Online communication on social media allows individuals to interact with a group of people who are like them (Oksanen et al., 2014). Social media use also reinforces homophily by the formation of identity bubbles (i.e., a tendency to relate oneself with online social cliques) (Kaakinen et al., 2018). In social media, individuals tend to communicate with people with similar interests and ideologies, which may relate to reduced diversity of social contacts and information diversity online (Kaakinen et al., 2018).

Researchers' increased interest in how individuals perceive their relationships with online networks and sharing information on social media (Kaakinen et al., 2018) can be examined through the often-used perspective of social identification, including social identity (Tajfel, 1978) and social categorization (Turner et al., 1987). This perspective is based on the notion that identifying membership in a social group affects our emotions and behavior as well as increases efficacy. The benefits of group identification have also been used to examine how individuals respond to threats and challenges, known as the "Social Cure" (Jetten et al., 2012).

In the context of COVID-19, social media use transformed young adults' means of interacting with others due to restrictions, which were especially disruptive to their social lives. In general, young adults' main purpose for social media use is communication, defined by contributing, sharing, consuming, searching for content, participating, and playing (Bolton et al., 2013). Social media use is driven by social enhancement (Ostic et al., 2021): by the need to interact with others (Palfrey & Gasser, 2008) and for socializing and experiencing a sense of community (Valkenburg et al., 2006), impacting psychological and emotional well-being (Bolton et al., 2013). It is also important for the development of one's identity by facilitating social interaction, forming and maintaining relationships, and sharing ideas (Kaakinen et al., 2018). In a life stage when young adults are expected to form relationships, in the context of the pandemic with nationwide restrictions on interaction with close friends, going online is basically the only way of maintaining close relationships.

Social media use also has other positive implications for its effect on *bridging* social capital (i.e., social interaction between individuals from different origins establishing connections through social networks) and *bonding* social capital (i.e., social and emotional support from close social networks), as mediating social media use and psychological well-being (Ostic et al., 2021). A heterogeneous network is especially important for social capital (Kim & Kim, 2017), as social interaction between individuals from different origins supports access to wider sources of information, interests, skills, and new resources which are shared with others (Ostic et al., 2021). The objective accuracy of opinions improves with exposure to a diverse (versus homogeneous) set of views (Gu et al., 2014). Social media, as a new communication channel, may build a sense of connectedness with relevant others, which reduces social isolation (Primack et al., 2017; Twenge & Campbell, 2019). This is especially evident with the increased use of sharing videos and images and other interpersonal connectivity provided by social media platforms (Whaite et al., 2018). During the pandemic, Ostic and colleagues (2021) show that social media use is a significant contributor to bonding social capital, in that social media allows young people to maintain their social relationships with close colleagues.

8.2.2 Young Adults' Financial Well-being

Financial well-being refers to a positive subjective assessment of one's current and future financial life (Brüggen et al., 2017) and is an essential component of holistic well-being (Jaggar & Navlakhi, 2021). For many young adults, the ability to manage one's money is an important component of adult status (Arnett, 2014; Serido et al., 2023). In accordance with the conceptualization of young adults' subjective financial well-being in previous research (Sorgente & Lanz, 2019; Sorgente et al., 2024b), the present study focuses on young adults' perception of their ability to manage material resources (i.e., money management). Because social media use and social relationships provide an opportunity for peer comparison and identity development (Thomas & Azmitia, 2014), we also investigate the relational dimension of financial well-being, where one's own financial condition is compared with peers' financial conditions (i.e., peer comparison).

In the economic crisis precipitated by COVID-19, individuals' financial resources were strained (Stevenson et al., 2022). Many young adults were already financially struggling prior to the pandemic, and their financial well-being continued to be eroded. In Finland, the context of this study, survey data on the impact of COVID-19 reported that 39 percent of young adults aged 18–25 were concerned about their personal financial situation (Ranta et al., 2020). Although objective financial circumstances such as income and debt contribute to financial well-being, how one feels about one's financial circumstances is of primary importance. Worldwide, many young adults looked to family for financial support (Sorgente et al., 2022), many of whom were also struggling financially (Stevenson et al., 2022). In these situations, financial strain is a shared experience among family members, including the psychological impact of financial challenges (Conger & Conger, 2002). Although the family as the primary social group and the means for collective psychological support is indeed important, for young adults the support given by peers is especially critical in this demanding life stage and particularly during macroeconomic shocks (Serido et al., 2014; Sorgente & Lanz, 2019). As a practical example, Hellerstein and Neumark (2020) found that in challenging financial times, individuals share information concerning informal insurance, risk-sharing, and potential job opportunities, especially among low-skilled and immigrant workers. In this study, we examine how young adults' homophilous social media use, that is, their tendency to communicate with people who share similar interests and ideologies, is associated with two dimensions of financial well-being: money management and peer comparison.

8.2.3 Financial Identity as a Moderating Force

Identity describes a self-regulating mechanism of development, providing a sense of personal control or self-agency (Crocetti, 2018). As a domain-specific component of identity, financial identity focuses on the development of financial functioning. In the present study, financial identity is operationalized by two dimensions of identity resolution: *exploration* of alternatives to identifications provided by significant others or society, and *commitment* to a set of values and beliefs (Marcia, 1966; Barber et al., 2011). When crossed, the two dimensions (i.e., exploration and commitment) yield four possible identity statuses: achievement, moratorium, foreclosure, and diffusion. *Achievement* involves commitment after exploration and is considered the most mature status. *Moratorium* is a period of active exploration and evaluation of alternatives. *Foreclosure* involves commitment without exploring, often to significant others' beliefs or values. *Diffusion* is the status lacking both firm commitments and the current experience of considering options. The four statuses are not mutually exclusive. Previous studies of young adults have found that a more active interest in finances (achieved and foreclosed financial identity statuses) is generally associated with more positive outcomes, while a lack of vested interest (i.e., moratorium and diffused financial identity statuses) is associated with more negative outcomes (Barber et al., 2011; Shim et al., 2013). Earlier studies also provide evidence that other external factors (e.g., family sociodemographic factors, education, parental interactions) influence financial identity status (Bosch et al., 2016; Vosylis & Erentaitė, 2019).

Although the formation of financial identity begins within the family when children are young, it continues to develop throughout life, influenced by the people and events that come and go (Gudmunson & Danes, 2011; Shim et al., 2013). We probe the possibility that one's financial identity may interact with homophilous social media use, altering its effect on financial well-being.

8.2.4 Current Study

As young adults engage in social media use, the development of their financial identity also occurs in this social interaction (Kaakinen et al., 2018; Serido et al., 2023). Social media use may, however, potentially contribute to the widening gap between groups, which restricts young adults' social networks and information seeking to content that reaffirms their own opinions while ignoring alternative views; precisely the definition of homophilous social media use.

On the other hand, social media may also be an important source of information or influence for coping with financial challenges and managing money in times of economic uncertainty.

In conclusion, this study examines sociodemographic differences in Finnish young adults' homophilous social media use and financial well-being, and the association between their homophilous social media use and two domains of financial well-being of direct relevance to young adults as they make their transition to financial independence during COVID-19. These are money management (i.e., self-perception of one's own financial skills) and peer comparison (i.e., how one is doing financially compared with peers) (Sorgente & Lanz, 2019). In addition, we examine the interaction of four measures of financial identity status with homophilous social media use and the effect on financial well-being.

8.3 DATA AND METHODOLOGY

8.3.1 Sample

We use empirical questionnaire data collected in Finland in December 2020 from 1,000 young adults aged 18 to 29 (M_{age}=23.53 (*SD*=3.44)). The survey included items to assess young adults' experiences and life situations during COVID-19, with a focus on their financial situation and digital skills. Although Finland has a Nordic welfare system that supports young adults' early financial independence, the growing global economic and social uncertainty, coupled with increasing social media use, poses challenges to their financial situations and agentic development (Wilska et al., 2021). Furthermore, during the time of data collection, Finland was once again experiencing a rise in COVID-19 infection rates after weeks of relative stability (Finnish Institute for Health and Welfare, 2021).

Participants responded to the survey via an online web panel. To represent the Finnish young adult population, stratified sampling was conducted based on quotas for age groups of 18–22, 23–25, and 26–29 years, gender, and region (large areas, NUTS2). About half (48.5 percent) were female, and 28 percent had completed a tertiary level degree. Regarding family financial status (SES), 22 percent rated their childhood family as somewhat disadvantaged or disadvantaged, 53 percent rated their family as neither disadvantaged nor wealthy (i.e., normal), and 25 percent rated their family as somewhat wealthy or wealthy.

8.3.2 Measures

8.3.2.1 Homophily

In assessing homophilous social media use, the homophily scales of the Identity Bubble Reinforcement Scale IBRS-6 and IBRS-9 were used: "In social media, I prefer interacting with people who are like me," "In social media, I prefer interacting with people who share similar interests with me," and "In social media, I prefer interacting with people who share my values" (Kaakinen et al., 2018; McCroskey et al., 1975; 2006). Higher scores refer to a higher preference for connecting with others who share the same views, with items having a scale ranging from 1 (does not describe me at all) to 7 (describes me perfectly) (α=0.739).

8.3.2.2 Financial well-being

In assessing financial well-being, two subscales from the Multidimensional Subjective Financial Well-being Scale were used: money management and peer comparison (Sorgente & Lanz, 2019). The measure was constructed specifically for assessing subjective financial well-being among young adults and is validated in cross-national studies (Iannello et al., 2021; Sorgente et al., 2024a). The money management dimension of financial well-being consists of four items (e.g., "I am satisfied with the way I manage my money") α(α=0.816), and the peer comparison consists of three items (e.g., "My financial situation is better than that of my peers"), measured on a Likert-type scale from 1 (absolutely false) to 5 (absolutely true) (α=0.738).

8.3.2.3 Financial identity

The participants indicated the extent to which each statement reflected their view of their financial identity using a 5-point scale (1=strongly disagree; 5=strongly agree). The four resulting identity statuses were: Achievement (e.g., "I have tried different ways to manage my personal finances and now I have a clear idea of what makes sense for me"; α=0.692); Foreclosure (e.g., "My parents know what's best for me in terms of how I should take care of my finances"; α=0.678); Moratorium (e.g., "There are so many different ways to manage money. I haven't decided which to follow, but I'm trying to figure it out"; α=0.616); and Diffusion (e.g., "I haven't really thought much about money management style. I'm not too concerned about credit ratings or paying bills"; α=0.685).

8.4 RESULTS

8.4.1 Descriptive Results

Mean and standard deviation for each key variable, as well as the correlation coefficients among key variables, are shown in Table 8.1.

8.4.1.1 Homophily

The average score of homophilous social media use was 3.90 (SD=1.29), slightly below the midpoint, indicating that the young adults in our sample were open to interacting with people who are not similar to them. One-way ANOVA for mean differences revealed significant differences in homophilous social media use by family SES (F(4, 985)=2.59, p=0.035). Young adults from the wealthiest families were more likely to interact with others who shared their interests, compared with those from somewhat disadvantaged to somewhat wealthy families (Figure 8.1). No differences were found by age, gender, or education status.

8.4.1.2 Financial well-being

The average score of money management was 3.29 (SD=0.87), above the midpoint, indicating that the young adults in our sample agreed that they were satisfied with their money management. Mean comparison analysis revealed

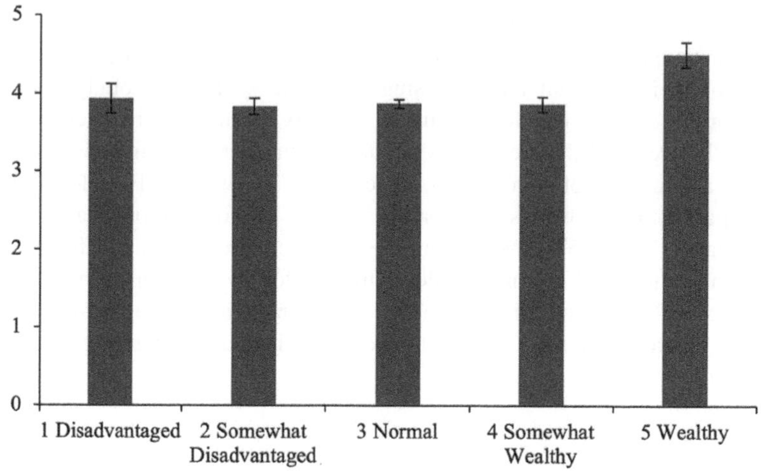

Figure 8.1 Homophilous social media use by family SES

Table 8.1 *Descriptives and correlations for key variables*

Descriptive	*M* (*SD*)	1	2	3	4	5	6	7	8	9	10
1. Diffusion	2.66 (0.91)	-	-	-	-	-	-	-	-	-	-
2. Moratorium	2.97 (0.83)	0.406***	-	-	-	-	-	-	-	-	-
3. Achievement	3.12 (0.85)	−0.053	−0.103**	-	-	-	-	-	-	-	-
4. Foreclosure	2.43 (0.93)	0.434***	0.325***	0.218***	-	-	-	-	-	-	-
5. Money management	3.29 (0.87)	−0.041	−0.219***	0.516***	0.126***	-	-	-	-	-	-
6. Peer comparison	3.10 (0.96)	−0.130***	−0.262***	0.230***	−0.025	0.431***	-	-	-	-	-
7. Homophily	3.90 (1.29)	0.153***	0.167***	0.087**	0.185***	0.015	−0.066*	-	-	-	-
8. Age	23.53 (3.44)	−0.059	−0.089**	0.137***	−0.191***	0.036	−0.04	−0.011	-	-	-
9. Gender	48.5 % F	−0.199***	−0.031	−0.109**	−0.318***	−0.094**	−0.032	−0.046	0.041	-	-
10. Education status	2.76 (1.14)	−0.100**	−0.053	0.207***	0.016	0.205***	0.158***	0.044	0.344***	0.002	-
11. Family SES	3.02 (0.87)	0.059	0.033	0.076*	0.188***	0.132***	0.178***	0.048	−0.136***	−0.082*	0.069*

Note: *** $p < 0.001$, ** $p < 0.01$, * $p < 0.05$.

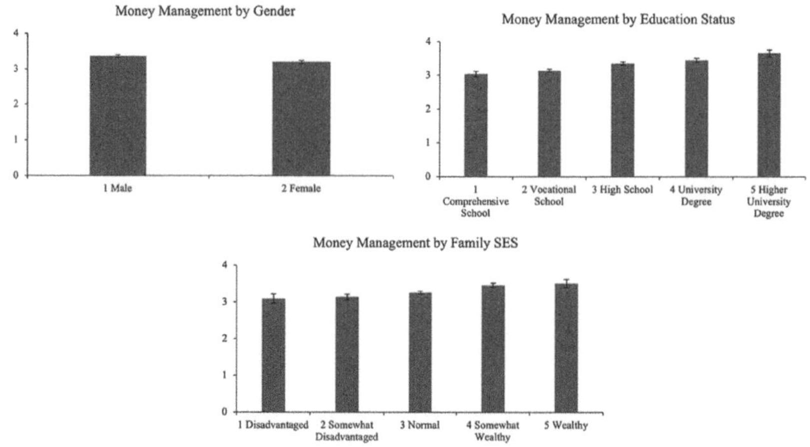

Figure 8.2 *Money management by gender, education status, and family SES*

significant differences in money management by gender, education status, and family SES (Figure 8.2).

The *t*-test for mean differences in money management across gender showed significant results. Male participants (*M*=3.37, *SD*=0.79, *SE*=0.03) reported higher money management well-being than female participants (*M*=3.20, *SD*=0.94, *SE*=0.04), *t*(945.096)=2.965, *p*=0.003.

One-way ANOVA for mean differences in money management across education statuses also showed a significant result (*F*(4, 991)=11.128, *p*<0.001). Young adults with at least a high school diploma (i.e., high school, undergraduate, and graduate degree) reported significantly higher levels of money management well-being than those who attended comprehensive and vocational schools. One-way ANOVA for mean differences in money management across family SES also showed a significant result (*F*(4, 985)=4.718, *p*=0.001). Young adults from somewhat wealthy families were more satisfied with their money management than those from somewhat disadvantaged families.

The average score of peer comparison was 3.10 (*SD*=0.96), close to the midpoint, indicating that the young adults in our sample perceived their financial situation as comparable to their peers. Mean comparison analysis revealed significant differences in peer comparison by education status and family SES (Figure 8.3).

One-way ANOVA for mean differences in peer comparison across education statuses showed a significant result (*F*(4, 991)=6.483, *p*<0.001). Young adults with an undergraduate or graduate degree were more likely to perceive

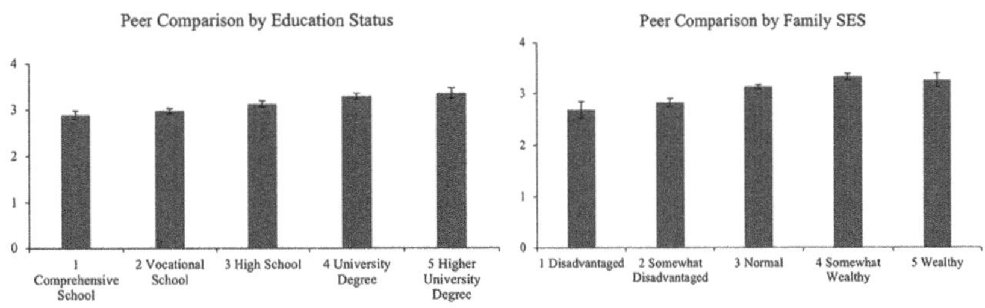

Figure 8.3 *Peer comparison by education status and family SES*

their own financial status as better than their peers, compared with those who attended comprehensive and vocational school. Similar analyses for mean differences in peer comparison across family SES showed a significant result ($F(4, 985)=9.223$, $p<0.001$). Young adults from normal (i.e., middle-class) and wealthier families were more likely to report a better-than-peer financial situation than those from disadvantaged families; young adults from normal (i.e., middle-class) and somewhat wealthy families were more likely to report a better-than-peer financial situation than those from somewhat disadvantaged families.

8.4.1.3 Financial identity

In our sample, three of the scores were slightly above the midpoint of the scale, with Achievement score being the highest (M=3.12; SD=0.853), followed by Moratorium (M=2.97; SD=0.826) and Diffusion (M=2.66; SD=0.906). The Foreclosure score was slightly below the midpoint (M=2.43; SD=2.97).

8.4.2 The Moderating Role of Financial Identity Status

We tested a series of eight hierarchical multiple regression models to estimate the strength of the associations of the four financial identity statuses with each of the two financial well-being outcomes (i.e., money management and peer comparison) (direct effects), and to detect a possible interaction of each financial identity status and homophily (moderating effect). In the first step of the model for each outcome (i.e., money management, peer comparison), we added sociodemographic variables as control variables. In the second step, we added homophily and one of the four financial identity status variables, and in the third step, we included the interaction term (for example, homophily × financial identity status). All analyses were conducted in SPSS.

8.4.2.1 Money management

Regarding direct effects after controlling for sociodemographic factors (i.e., age, gender, education status, family SES), we found that homophilous social media use was not related to money management. Achieved and foreclosure identity were positively associated with money management ($\beta_{achieved}$=0.493, p<0.001 and $\beta_{foreclosed}$=0.082, p=0.017 respectively), and moratorium identity was negatively associated with money management ($\beta_{moratorium}$=−0.227, p<0.001). There were no significant direct effects of diffused identity. These results were consistent with previous literature (Barber et al., 2011; Shim et al., 2013).

We did find a significant interaction: diffused identity significantly moderated the association between homophilous social media use and money management, $\beta_{interaction}$=0.085, p=0.007, R^2=0.065. Specifically, young adults who registered low homophilous social media use and low diffused identity status registered higher levels of money management financial well-being compared with young adults who registered low homophilous social media use and high diffused identity. When young adults registered high homophilous social media use, there was no difference in money management financial well-being between those with high levels of diffused identity versus those with low diffused identity (Figure 8.4).

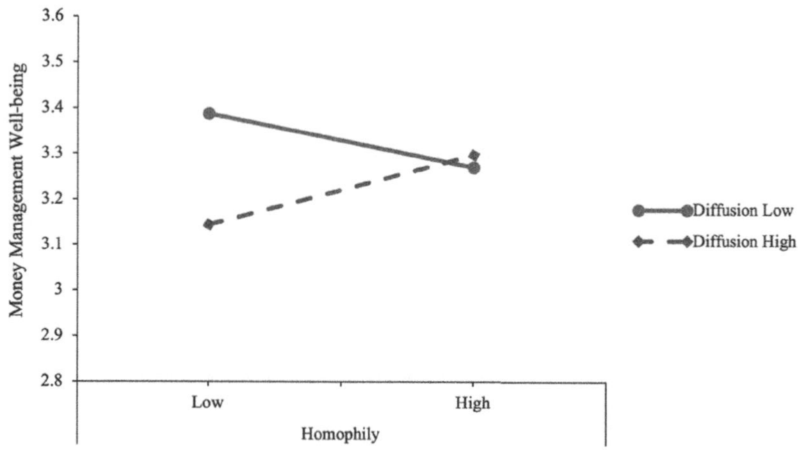

Figure 8.4 Effect of homophily on money management well-being moderated by diffusion

8.4.2.2 Peer comparison

Regarding direct effects after controlling for sociodemographic factors (i.e., age, gender, education status, family SES), homophilous social media use remained significantly negatively related to peer comparison even after adding three of the financial identity status predictors: achieved identity ($\beta_{homophily}$=–0.099, p=0.001), foreclosure identity ($\beta_{homophily}$=–0.080, p=0.012), or diffused identity ($\beta_{homophily}$=–0.065, p=0.036). In addition, achieved identity was positively associated with peer comparison ($\beta_{achieved}$=0.212, p<0.001), and moratorium and diffused identity were negatively associated with peer comparison ($\beta_{moratorium}$=–0.254, p<0.001 and $\beta_{diffused}$=–0.120, p<0.001 respectively). There were no significant direct effects of foreclosure identity. Although there are no empirical studies examining the associations between financial identity and peer financial comparison, the results align with the conceptualization of financial identity as a lack of vested interest in finances (i.e., diffused, moratorium) versus an active interest in finances (Barber et al., 2011; Shim et al., 2013).

Despite a lack of a significant direct effect of foreclosure identity, we found a significant interaction: foreclosure identity significantly moderated the association between homophily and peer comparison, $\beta_{interaction}$=–0.068, p=0.030, R^2=0.075. Specifically, young adults who registered high homophilous social media use and low foreclosure identity status registered higher levels of peer comparison compared with young adults who registered high homophilous social media use and high foreclosed identity. When young adults registered low homophilous social media use, there was no difference in peer comparison between those with high levels of foreclosure identity versus those with low foreclosure identity (Figure 8.5).

8.5 DISCUSSION

Previous studies focusing on young adults in a digital world have found mixed results on the effects of social media use on well-being (Berryman et al., 2018; Cho et al., 2023; Lee et al., 2022; Ostic et al., 2021; Twenge & Campbell, 2019). In the context of economic disruptions and lockdown restrictions due to COVID-19, we investigated differences in young adults' homophilous social media use and financial well-being by sociodemographic factors and the association between homophilous social media use and financial well-being among Finnish young adults, moderated by their financial identity.

Consistent with the extant literature (Danes & Brewton, 2014; Luhr, 2018; LeBaron & Kelley, 2020), we find that sociodemographic factors play an important role in young adults' social media use and financial well-being. Specifically, young adults from wealthier families engaged in more homophilous social media use, reported higher money management well-being, and were

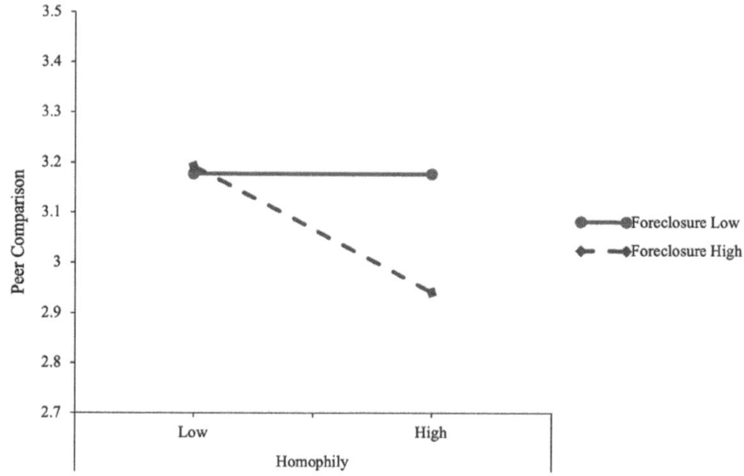

Figure 8.5 *Effect of homophily on peer comparison moderated by foreclosure*

better off financially compared with their peers. Higher-educated young adults reported higher money management well-being and were better off financially compared with their peers. Males reported higher money management skills.

Our results on homophilous social media use reflect the likelihood that young adults were using social media for different reasons (e.g., support-seeking individuals were connecting with others with similar views (Cho et al., 2023; Lisitsa et al., 2020)) versus information-seeking individuals being more open to diverse views (Kapoor et al., 2018; Kim et al., 2011). Our key finding provides evidence of conditional benefits: (1) more diverse social media use (low homophily) when young adults expressed some interest in finances (low diffusion) was associated with higher money management well-being; (2) less diverse social media use (high homophily) when young adults expressed some openness to exploring financial alternatives (low foreclosure) was associated with more peer comparison well-being.

8.5.1 Benefits and Detriments of Homophilous Social Media Use within the Financial Domain

Some researchers have argued that it may not be the time spent online but rather the quality and content of these online activities that matter for well-being (Berryman et al., 2018). In this sense, our findings on the conditional effects of homophilous social media use extend that line of thought to consider how multiple self-views evolve and interact during young adulthood (Crocetti, 2018; Negru-Subtirica et al., 2017). During societal disruption, such as the pandemic lockdown, it makes sense that young adults would seek to interact with like-minded individuals to maintain relationships and a sense of community (Valkenburg et al., 2006) and thus promote psychological well-being (Bolton et al., 2013). Yet, in addition to social interaction (Kaakinen et al., 2018), young adults also need to develop financial decision-making skills in an era of increasing responsibility over one's financial future (Serido et al., 2014).

From this perspective, higher homophilous social media use appears to be beneficial (associated with feeling financially better off than peers) for individuals who are still exploring their identity (low identity foreclosure). Higher homophilous social networks may offer a safe space for discussion. We speculate that low identity foreclosure (still exploring) may encourage young adults to discuss the financial challenges they encounter with others who share similar interests. In turn, young adults may be better able to put their own situation in a wider context and see it in a more positive light. In contrast, more open (lower homophilous) social networks may provide access to new information that challenges personal views and promotes growth (Gu et al., 2014; Ostic et al., 2021). Applied to our findings, young adults who previously expressed little interest in finances may benefit from interactions with a more diverse social network. Hearing about the widespread financial challenges facing their peers, they may now see a reason to pay more attention to finances.

8.5.2 Practical Implications

Despite the potential challenges of social media use, its use among young adults is a reality of contemporary life. As educators, researchers, and parents, we must focus on what can and should be done for the next generation in this regard. Knowing that young adults are increasingly dependent on information gained online (Kaakinen et al., 2018), one important area the OECD International Network for Financial Education (INFE) calls for is further exploration in providing strategies for identifying credible resources with digital financial literacy and the need to study the implementation of these strategies at an international level (OECD, 2018).

We also need to acknowledge the importance of role models in promoting decision-making skills, particularly in building financial capability as young adults seek advice and information. Social media can efficiently disseminate information to young adults in the community, an excellent example being healthcare information provision (Bolton et al., 2013). The significant positive impact of homophily on the likelihood of following financial advice was identified by Stolper and Walter (2019) among clients and their bank advisors. In the context of the current state of the banking industry, with digitalization and "robo-advice," the value of human interaction and communication becomes of particular importance and therefore targeted client–advisor pairings could be vital for information transmission due to mutual understanding and shared characteristics and interests.

8.5.3 Avenues for Future Research

This study investigated young adults' homophilous social media use and financial well-being as they make their transition to financial independence during the exceptional time of the COVID-19 pandemic. Although the family as the primary social group and the means for collective psychological support is indeed important for young adults, the support given by peers via social media is also critical in this demanding life stage, particularly during macroeconomic shocks (Serido et al., 2014; Sorgente & Lanz, 2019). While the long-term effects of the pandemic on young adults' social media use and financial well-being remain unforeseen and an important avenue for future research, young adults' financial identity, the self-perceived view of oneself in the financial domain (Serido et al., 2023), may serve as an internal compass to navigate the financial challenges a young adult may encounter.

AUTHORS' NOTE

This work was supported by the Strategic Research Council (SRC) at the Academy of Finland (#352545).

REFERENCES

Arnett, J. J. (2014). *Emerging adulthood: The winding road from the late teens through the twenties* (2nd ed.). Oxford University Press. https://doi.org/10.1093/acprof:oso/9780199929382.001.0001.

Barber, B. L., Card, N. A., Serido, J., & Shim, S. (2011, November). *Financial identity of United States and Australian college students*. Paper presented at the National Council of Family Relations Conference, Orlando, FL.

Berryman, B. C., Ferguson, C. J., & Negy, C. (2018). Social media use and mental health among young adults. *Psychiatric Quarterly*, *89*, 307–314. https://doi.org/10.1007/s11126-017-9535-6.

Bolton, R., Parasuraman, A. P., Hoefnagels, A., Migchels, N., Kabadayi, S., Gruber, T., Komarova, Y., & Solnet, D. (2013). Understanding Gen Y and their use of social media: A review and research agenda. *Journal of Service Management*, *24*, 245–267. https://doi.org/10.1108/09564231311326987.

Bosch, L. A., Serido, J., Card, N., Shim, S., & Barber, B. L. (2016). Predictors of financial identity development in emerging adulthood. *Emerging Adulthood*, *4*(6), 417–426. https://doi.org/10.1177/2167696816631845.

Brüggen, E. C., Hogreve, J., Holmlund, M., Kabadayi, S., & Löfgren, M. (2017). Financial well-being: A conceptualization and research agenda. *Journal of Business Research*, *79*, 228–237. https://doi.org/10.1016/j.jbusr es.2017.03.013.

Cho, H., Li, P., Ngien, A., Tan, M. G., Chen, A., & Nekmat, E. (2023). The bright and dark sides of social media use during COVID-19 lockdown: Contrasting social media effects through social liability vs. social support. *Computers in Human Behavior*, *146*, 107795. https://doi.org/10.1016/j.chb.2023.107795.

Conger, R. D., & Conger, K. J. (2002). Resilience in Midwestern families: Selected findings from the first decade of a prospective, longitudinal study. *Journal of Marriage and Family*, *64*(2), 361–373. https://doi.org/10.1111/j.1741–3737.2002.00361.x.

Crocetti, E. (2018). Identity dynamics in adolescence: Processes, antecedents, and consequences. *European Journal of Developmental Psychology*, *15*(1), 11–23. https://doi.org/10.1080/17405629.2017.1405578.

Danes, S. M., & Brewton, K. E. (2014). The role of learning context in high school students' financial knowledge and behavior acquisition. *Journal of Family and Economic Issues*, *35*(1), 81–94. https://doi.org/10.1007/s10834-013-9351-6.

Delbosc, A., & McCarthy, L. (2021). Pushed back, pulled forward: Exploring the impact of COVID-19 on young adults' life plans and future mobility. *Transport Policy*, *107*, 43–51.

Finnish Institute for Health and Welfare (2021). Number of COVID-19 cases on the rise again in Finland after several weeks of stability. https://thl.fi/en/-/number-of-covid-19-cases-on-the-rise-again-in-finland-after-several-weeks-of-stability.

Grzegorczyk, M., & Wolff, G. (2020). The scarring effect of COVID-19: Youth unemployment in Europe. *Bruegel Blog*. https://www.bruegel.org/2020/11/the-scarring-effect-of-covid-19-youth-unemployment-in-europe/.

Gu, B. Konana, P., Raghunathan, R., Chen, H. M. (2014). Research note: The allure of homophily in social media: Evidence from investor responses on virtual communities. *Information Systems Research*, *25*(3), 604–617.

Gudmunson, C. G., & Danes, S. M. (2011). Family financial socialization: Theory and critical review. *Journal of Family and Economic Issues*, *32*, 644–667. https://.doi.org/10.1007/s10834-011-9275-y.

Hellerstein, J. K., & Neumark, D. (2020). Social capital, networks, and economic wellbeing. *Future of Children*, *30*(1), 127–152.

Iannello, P., Sorgente, A., Lanz, M., & Antonietti, A. (2021). Financial well-being and its relationship with subjective and psychological well-being among emerging adults: Testing the moderating effect of individual differences. *Journal of Happiness Studies*, *22*(3), 1385–1411. https://doi.org/10.1007/s10902-020-00277-x.

Jaggar, S., & Navlakhi, L. (2021). Financial wellbeing: The missing piece in holistic wellbeing. *NHRD Network Journal*, *14*(1), 83–94. https://doi.org/10.1177/2631454120980600.

Jetten, J., Haslam, C., & Haslam, S. (2012). *The social cure: Identity health and wellbeing*. Routledge.

Kaakinen, M., Sirola, A., Savolainen, I., & Oksanen, A. (2018). Shared identity and shared information in social media: Development and validation of the identity bubble reinforcement scale. *Media Psychology*, *23*(1), 25–51. https://doi.org/10.1080/15213269.2018.1544910.

Kapoor, K. K., Tamilmani, K., Rana, N. P., Pushp, P., Dwivedi, Y. K., & Nerur, S. (2018). Advances in social media research: Past, present and future. *Information Systems Frontiers*, *20*, 531–558. https://doi.org/10.1007/s10796-017-9810-y.

Kim, B., & Kim, Y. (2017). College students' social media use and communication network heterogeneity: Implications for social capital and subjective well-being. *Computers in Human Behavior, 73*, 620–628. https://doi.org/10.1016/j.chb.2017.03.033.

Kim, K.-S., Yoo-Lee, E., & Sei-Ching, J. S. (2011). Social media as information source: Undergraduates' use and evaluation behavior. *Proceedings of the American Society for Information Science and Technology, 48*, 1–3. https://doi.org/10.1002/meet.2011.14504801283.

Laakasuo, M., Rotkirch, A., van Duijn, M., Berg, V., Jokela, M., David-Barrett, T., et al. (2020). Homophily in personality enhances group success among real-life friends. *Frontiers in Psychology, 11*, 710. https://doi.org/10.3389/fpsyg.2020.00710.

LeBaron, A. B., & Kelley, H. H. (2020). Financial socialization: A decade in review. *Journal of Family and Economic Issues, 42*(Suppl1), 195–206. https://doi.org/10.1007/s10834-020-09736-2.

Lee, Y., Jeon, Y. J., Kang, S., Shin, J. I., Jung, Y.-C., & Jung, S. J. (2022). Social media use and mental health during the COVID-19 pandemic in young adults: A meta-analysis of 14 cross-sectional studies. *BMC Public Health, 22*(1), 995. https://doi.org/10.1186/s12889-022-13409-0.

Lisitsa, E., Benjamin, K. S., Chun, S. K., Skalisky, J., Hammond, L. E., & Mezulis, A. H. (2020). Loneliness among young adults during Covid-19 pandemic: The mediational roles of social media use and social support seeking. *Journal of Social and Clinical Psychology, 39*(8), 708–726. https://doi.org/10.1521/jscp.2020.39.8.708.

Luhr, S. (2018). How social class shapes adolescent financial socialization: Understanding difference in the transition to adulthood. *Journal of Family and Economic Issues*, 39(3), 457–473. https://doi.org/10.1007/s10834-018-9573-8.

Marcia, J. E. (1966). Development and validation of egoidentity status. *Journal of Personality and Social Psychology, 3*(5), 551–558. https://doi.org/10.1037/h0023281.

McCroskey, J. C., Richmond, V. P., & Daly, J. A. (1975). The development of a measure of perceived homophily in interpersonal communication. *Human Communication Research, 1*(4), 323–332. https://doi.org/10.1111/hcre.1975.1.issue-4.

McCroskey, L. L., McCroskey, J. C., & Richmond, V. P. (2006). Analysis and improvement of the measurement of interpersonal attraction and homophily. *Communication Quarterly, 54*(1), 1–31. https://doi.org/10.1080/01463370500270322.

McPherson, M., Smith-Lovin, L., & Cook, J. M. (2001). Birds of a feather: Homophily in social networks. *Annual Review of Sociology, 27*(1), 415–444. https://doi.org/10.1146/annurev.soc.27.1.415.

Negru-Subtirica, O., Pop, E. I., & Crocetti, E. (2017). A longitudinal integration of identity styles and educational identity processes in adolescence. *Developmental Psychology, 53*(11), 2127–2138. https://doi.org/10.1037/dev0000325.

OECD (2018). *G20/OECD INFE policy guidance on digitalisation and financial literacy.* https://www.oecd.org/g20/summits/buenos-aires/G20-OECD-INFE-Policy-Guidance-Digitalisation-Financial-Literacy-2018.pdf.

Oksanen, A., Hawdon, J., & Räsänen, P. (2014). Glamorizing rampage online: School shooting fan communities on YouTube. *Technology in Society, 39*, 55–67. https://doi.org/10.1016/j.techsoc.2014.08.001.

Ostic, D., Qalati, S. A., Barbosa, B., Shah, S. M. M., Galvan, V. E, Herzallah, A. M., & Liu, F. (2021). Effects of social media use on psychological well-being: A mediated model. *Frontiers in Psychology, 12*, 678766. https://doi.org/10.3389/fpsyg.2021.678766.

Palfrey, J., & Gasser, U. (2008). *Born digital: Understanding the first generation of digital natives.* Basic Books.

Primack, B. A., Shensa, A., Sidani, J. E., Whaite, E. O., Yi Lin, L., Rosen, D., et al. (2017). Social media use and perceived social isolation among young adults in the US. *American Journal of Preventative Medicine*, *53*, 1–8. https://doi.org/10.1016/j.amepre.2017.01.010.

Ranta, M., Silinskas, G., & Wilska, T-A. (2020). Young adults' personal concerns during the COVID-19 pandemic in Finland: An issue for social concern. *International Journal of Sociology and Social Policy*, *40*(9/10), 1201–1219. https://doi.org/10.1108/IJSSP-07-2020-0267.

Serido, J., Shim, S., Xiao, J. J., Card, N., & Tang, C. (2014). Financial adaptation among college students: Helping students cope with financial strain. *Journal of College Student Development*, *55*(3), 310–316. http://dx.doi.org/10.1353/csd.2014.0032.

Serido, J., Sorgente, A., Lanz, M., & Shim, S. (2023). Becoming self-sufficient: A longitudinal person-centered analysis of financial identity and adult status during emerging adulthood. *Applied Developmental Science*, *27*(2), 99–114. https://doi.org/10.1080/10888691.2022.2035224.

Shanahan, L., Steinhoff, A., Bechtiger, L., Murray, A. L., Nivette, A., Hepp, U., Ribeaud, D., & Eisner, M. (2020). Emotional distress in young adults during the COVID-19 pandemic: Evidence of risk and resilience from a longitudinal cohort study. *Psychological Medicine*, *52*(5), 824–833. https://doi.org/10.1017/S003329172000241X.

Shim, S., Serido, J., Bosch, L., & Tang, C. (2013). Financial identity-processing styles among young adults: A longitudinal study of socialization factors and consequences for financial capabilities. *Journal of Consumer Affairs*, *47*(1), 128–152. https://doi.org/10.1111/joca.12002.

Sorgente, A., Atay, B., Aubrey, M., Bhatia, S., Crespo, C., Fonseca, G., et al. (2024a). One (financial well-being) model fits all? Testing the Multidimensional Subjective Financial Well-being Scale across nine countries. *Journal of Happiness Studies*, *25*(13), 1–26. https://doi.org/10.1007/s10902-024-00708-z.

Sorgente, A., Fonseca, G., Lep, Z., Li, L., Serido, J., Rimantas, V., et al. (2022). Profiles of emerging adults' resilience facing the negative impact of COVID-19 across six countries. *Current Psychology*, 1–13. https://doi.org/10.1007/s12144-022-03658-y

Sorgente, A., & Lanz, M. (2019). The multidimensional subjective financial well-being scale for emerging adults: Development and validation studies. *International Journal of Behavioral Development*, *43*(5), 466–478. https://doi.org/10.1177/0165025419851859.

Sorgente, A., Serido, J., Lanz, M., & Shim, S. (2024b). Family financial socialization during emerging adulthood: Insights from a cross-lagged panel model. *Journal of Family Psychology*, *38*(1), 161–173. https://dx.doi.org/10.1037/fam0001139.

Statista (2023). *Digital 2023: October Global Statshot Report.* https://datareportal.com/reports/digital-2023-october-global-statshot (retrieved 9 May 2024).

Stevenson, C., Wakefield, J. R. H., Bowe, M., Kellezi, B., Jones, B., & McNamara, N. (2022). Weathering the economic storm together: Family identification predicts future well-being during COVID-19 via enhanced financial resilience. *Journal of Family Psychology*, *36*(3), 337–345. https://doi.org/10.1037/fam0000951.

Stolper, O., & Walter, A. (2019). Birds of a feather: The impact of homophily on the propensity to follow financial advice. *The Review of Financial Studies*, *32*(2), 524–563. https://doi.org/10.1093/rfs/hhy082.

Tajfel, H. (1978). The achievement of inter-group differentiation. In H. Tajfel (Ed.), *Differentiation between social groups* (pp. 77–100). Academic Press.
Thomas, V., & Azmitia, M. (2014). Does class matter? The centrality and meaning of social class identity in emerging adulthood. *Identity, 14*, 195–213. https://doi.org/10.1080/15283488.2014.921171.
Turner, J. C., Hogg, M., Oakes, P., Reicher, S., & Wetherell, M. (1987). *Rediscovering the social group: A self-categorization theory*. Basil Blackwell.
Twenge, J. M., & Campbell, W. K. (2019). Media use is linked to lower psychological well-being: Evidence from three datasets. *Psychiatric Quarterly, 90*, 311–331. https://doi.org/10.1007/s11126-019-09630-7.
Valkenburg, P., Jochen, P., & Schouten, A. (2006). Friend networking sites and their relationship to adolescents' well-being and social self-esteem. *Cyberpsychology & Behavior: The Impact of the Internet, Multimedia and Virtual Reality on Behavior and Society, 9*, 584–590. https://doi.org/10.1089/cpb.2006.9.584.
Vehkalahti, K., Armila, P., & Sivenius, A. (2021). Emerging adulthood in the time of pandemic: The COVID-19 crisis in the lives of rural young adults in Finland. *Young, 29*(4), 399–416. https://doi.org/10.1177/11033088211026502.
Vosylis, R., & Erentaitė, R. (2019). Linking family financial socialization with its proximal and distal outcomes: Which socialization dimensions matter most for emerging adults' financial identity, financial behaviors, and financial anxiety? *Emerging Adulthood, 8*(6), 464–475. https://doi.org/10.1177/2167696819856763.
Whaite, E. O., Shensa, A., Sidani, J. E., Colditz, J. B., & Primack, B. A. (2018). Social media use, personality characteristics, and social isolation among young adults in the United States. *Personality and Individual Differences, 124*, 45–50. https://doi.org/10.1016/j.paid.2017.10.030.
Wilska, T.-A., Sirola, A., Nuckols, J., & Nyrhinen. J. (2021). A year of COVID-19 in three countries – a study on the effects of the COVID-19 pandemic on everyday life, consumption and digital behaviour in Finland, Sweden, and Great Britain. *YFI julkaisuja*. Jyväskylä: Jyväskylä University Printing House.
World Bank, The (2023). Collapse and recovery: How COVID-19 eroded human capital and what to do about it. https://www.worldbank.org/en/publication/human-capital/publication/collapse-recovery-how-covid-19-eroded-human-capital-and-what-to-do-about-it.

9. Digital agency and vulnerability: a study on young adult consumers in personalised advertising ecosystems

Mikko Laamanen, Erlend Kok, Arne Dulsrud and Dag Slettemeås

9.1 INTRODUCTION

The purpose of this chapter is to explore the complexity of the processes of personalised advertising based on personal data and behavioural targeting, and how consumers consider themselves both 'seen' and rendered vulnerable. We build on previous work on how consumers understand the connection between their data and the advertising they see (Ruckenstein & Granroth, 2019; Sörum & Fuentes, 2023) and their capacity to address becoming targeted by advertisers who draw on their personal and behavioural data (Draper & Turow, 2019). Importantly, as Lupton (2021), and Sörum and Fuentes (2023) highlight, consumers may not only be apathetic victims to data use and profiling, but they also actively engage in agentic reimagining of their data and algorithmic manipulations. Wishing to highlight the 'privacy paradox' (Barnes, 2006) in action, we engaged with young consumers' perceived agency in digital contexts in Norway[1] but also want to account for how personalised advertising may amplify their experiences of vulnerability.

Central to our endeavour are the concepts of agency and vulnerability. Agency relates to who has access to the consumer's personal data, that is, 'those who collect, store, and mine large quantities of data' or 'those whom data collection targets' (Andrejevic, 2014, p. 1673). Vulnerability is defined as the susceptibility to influence that makes one come out worse in consumption than would be normal (OECD, 2023). Experienced vulnerabilities relate to consumers' concern that lacking data privacy leads to a susceptibility to harm (Cartwright, 2014), including 'feelings of distress, confusion, and alienation as well as the more direct consequences of identity theft and grift' (Carter & Eger, 2021, p. 374). Empirically, we aim to answer the question of how agency and vulnerability are experienced in the advertising ecosystem, and study the discursive processes of advertising market participants, advertising market stakeholders, and consumers, paying special attention to young adults. Over 90 per cent of Norwegian young adults use social media (nearly) every day (Statistics Norway, 2023) – this group is widely considered to be confronted with and potentially vulnerable to advertising and other forms of social media influence that draw from their personal and behavioural data (Batat, 2010; O'Day & Heimberg, 2021; Rounsefell et al., 2020).

Our findings contribute to more knowledge about the users' life situations which, we presume, can be linked back to the use of different types of personal data used in advertising. Our findings also show what Lupton (2021, p. 16) calls 'pragmatic acceptance of the inevitability of practices such as data profiling and targeted advertising'. Importantly, the vulnerability discussed by the industry representatives focused similarly on the inevitability that all consumers are inherently vulnerable to data-based advertising practices and should be aware as well as protected against the increasing harmfulness of these practices.

9.2 THEORETICAL FRAMEWORK

Digital advertising refers to commercial messages delivered to target groups through various web-based media with various approaches and categories, such as search (e.g., SEO, 'retargeting'), social media, 'content' (i.e., messages that mix entertainment, information, and advertising in 'advertainment' or 'advertorials'), and influencer marketing (European Commission, 2022; Hackley & Hackley, 2021). The digital transition of the last decades has changed the advertising market in various ways (Hackley & Hackley, 2021; Jenkins, 2006; Lee & Cho, 2020; Lynch, 2018). First, traditional media – newspapers, magazines, and television – have experienced a drastic reduction in audiences: this 'nichefication' of media consumption is described as 'media fragmentation' and 'audience fragmentation'. Second, a duopoly of Alphabet and Meta dominates the global digital advertising market, both in terms of data analytics and media space, accounting for over half of the global ad spend. Third, programmatic advertising using automated big data systems has changed how advertising space is sold based on the behavioural insight on the users (Samuel et al., 2021; Smit et al., 2014). Fourth, a greater focus on 'owned media' (brand-controlled social media outlets) and 'earned media' (commonly social media exposure through sharing and liking) instead of paying for advertising as 'bought media' (Colicev et al., 2018) contributes to an increase in social media communication campaigns, more user-generated content, and a more extensive influencer culture. Fifth, an amplified use of advertising avoidance technologies – consumers' use of 'ad-blockers' – coincides with mobile phones dominating as an entry point to media consumption ('media convergence').

Digital marketing flourishes because of the various potential data points that can be collected and used for audience profiling and targeting. Compared with marketing through traditional (legacy) media, digital is perceived more accessible, immediate, and social without space and time limitations. Advertisers are often attracted by the perceived effectiveness of digital solutions measured in cost, reach, and conversion; but above all, the increased ability to predict (and prefigure) needs with greater precision to trigger action (e.g., purchases). As traditional media struggle to achieve anything close to the same audience reach they once had, digital advertising has become the mainstay of advertising expenditure (Dentsu, 2023; Lee & Cho, 2020). Real-time data and analysis allow for the personalisation of advertising, that is, the collection, use, and storage of recognisable, personal aspects in the content information of the recipients (Hackley & Hackley, 2021; Smit et al., 2014).

Consumers engage with a variety of platforms and, in the process, become exposed to numerous advertisements a day; they may experience personalised advertising as more relevant and useful, truthful to their individual needs and interests (Ruckenstein & Granroth, 2019), as this creates personalised advertisements based upon information available about the user (Farahat & Bailey, 2012). Yet, the ways in which consumers are reached in the digital space are increasingly being questioned. The interests of consumers and various advertising ecosystem actors clash, particularly in relation to safeguards to consumer privacy, informed consent, and amplified vulnerabilities. Previous research has shown how a link between sensitive personal data (e.g., ethnicity, religious and political views, marital status, gender, sexual orientation) and the likelihood of consumption of alcohol, cigarettes, and drugs can be made based on online behaviour ('likes', Spiekermann & Christl, 2016). Behavioural targeting in programmatic advertising plays on similar extrapolation.

Thus, advertising personalisation depends on data, which is the currency and driver of digital capitalism (Fuchs, 2017). While data are argued to be essential to create products that meet consumer needs, the processes of collecting and analysing consumer data are increasingly called into question. Concepts such as data capitalism (West, 2019), surveillance capitalism (Zuboff, 2015) and dataveillance (Ruckenstein & Granroth, 2019) point to the underlying logic of accumulation in digital data collection. Milyaeva and Neyland (2016, p. 230) elaborate on how data can 'flow freely as a raw material … but its mining and processing turn it into a tradable commodity', and as West (2019) describes, the digital ecosystem has transformed from a sales channel to a medium for harvesting user data used for sales. In particular, data brokers and data platforms have become central to extractive processes. The legitimacy of data capitalism has been called into question due to the dubious practices that the middlemen of advertising technology, 'ad tech', engage in to sell programmatic advertising (ICO, 2019). The logic of accumulation draws on the data points individuals leave behind in the online space – the value of these data are integrated into the networks that bond the collection, analysis, and commercialisation of data, as well as the surveillance and manipulation needed to maintain the data ecosystem. In our empirical study, we pose the question of how agency and vulnerability are experienced in the advertising ecosystem. Beyond young adult consumers, we attend to the institutional environment including advertisers, regulators, and NGOs. We approach the ways in which Norwegian digital advertising market actors understand and position themselves in relation to data-driven behavioural targeting and its moral contradictions related to vulnerability.

9.3 METHODS AND SAMPLE

This chapter draws on research in the Norwegian digital advertising ecosystem, where we focused on the understandings that the various participating actors have of the process of personalisation and targeting of advertising, and its impact on vulnerability. First, setting the context of the institutional understanding of data practices in digital advertising, we then want to illustrate how the consumers experience data-driven targeting and its personal implications for them. While statistics and ecosystem structures are specific to the Norwegian context, the actor roles and processes described apply more universally.

9.3.1 Research Context

The total turnover of the Norwegian advertising market is estimated at NOK 25.2 billion (approximately €2.2 billion) with an increase of 2.4 per cent in 2024 according to the Norwegian Advertisers' Association ANFO (2023). Norwegian media consumption, and hence also the advertising market, is mostly digital. Of the Norwegian population, 93 per cent aged 9–79 use the internet daily, for approximately 218 minutes in total. A further 64 per cent read online newspapers and 51 per cent watch video and streaming media (Schiro, 2023). Social media is used daily by 78 per cent of Norwegian consumers, with dominant platforms being Facebook (67 per cent), Instagram (43 per cent), and Snapchat (53 per cent) (Medienorge, 2024). Consequentially, advertisers are making most investment in email/SMS marketing, searching, social media, and online video (ANFO, 2023).

To target digital advertising, most advertisers purchase data and analysis services from third-party providers. Large global platforms such as Meta, Google, and TikTok play important roles in the Norwegian ad tech industry as media and third-party data suppliers. These actors combine various data sources to be able to provide a nuanced and accurate picture of the individual user (Slettemeås & Storm-Mathiesen, 2021) and play a critical role in 'real-time bidding' processes where advertising space is sold to 'eyeballs' in a matter of milliseconds (Datatilsynet, 2015). Although the user may experience some benefits (Ruckenstein & Granroth, 2019), there is still a question of whether there is real consent and transparency in the data collection and sharing practices (ICO, 2019). As West (2019, p. 37) argues 'the capacity to collect and render intelligible the massive amount of digital traces produced by the population is only available to a select few', and this illustrates the concentration of power in certain actors. In effect, behavioural targeting and programmatic advertising are 'black boxes' because:

> neither the publisher (e.g. the owner of a website or app) nor the advertiser (e.g. the owner of the brand that is promoted) choose which ads to show and where to display them. (Forbrukerrådet, 2020, p. 8)

Some national media actors, such as Schibsted, represent a blurred distinction between advertiser and media supplier. They both collect, process, and analyse personal data through their various apps and offer advertising space through their ownership of large online media and marketplace services; they claim to reach 92 per cent of the national media market (Schiro, 2023).

From a regulatory point of view, control over how third parties use data beyond consumer consent is a prevailing concern. In recent years, the particular focus in Norway has been on consumer profiling and behavioural, 'surveillance-based' advertising (Forbrukerrådet, 2020, 2021). Consequently, the various participants in the advertising value chain are woven into a regulatory-political framework and surrounded by regulatory institutions that protect consumers and by interest organisations that look after the rights and interests of the population and/or specific groups.

To illustrate the dynamic set-up of the market, we conducted a two-part qualitative study in which we divided the participants in the Norwegian advertising market into two categories – Institutional Actors and Consumers. We explain our empirical approach in detail below.

9.3.2 Institutional Actors

To construct an institutional understanding of personalised advertising, we interviewed a total of 15 people from 11 different organisations representing different market positions of advertisers, advertising space suppliers, interest organisations and market regulators (see Table 9.1). The interviews took place between March and September 2023 and lasted approximately 60

Table 9.1 Informant characteristics – institutional actors

Organisation	Role and relation to personalised advertising
The Advertisers' Association (Annonsørforeningen, ANFO)	*Advertiser*: Representative for Norwegian advertisers in terms of the effectiveness and legality of advertising
Schibsted (SBT)	*Supplier*: The largest Norwegian publicist and seller in the ad tech market
Norwegian Media Businesses' Association (Mediebedriftenes Landsforening, MBL)	*Supplier*: The representative organisation for Norwegian media companies with a particular focus on the link to economic health for free media
Kobler (KB)	*Supplier*: Kobler offers contextual advertising, that is, commercial messages linked to editorial content, without the use of personal data
The Norwegian Data Protection Authority (Datatilsynet, DT)	*Regulator*: Focuses on data use in advertising (including consent mechanisms and fair data collection)
The Equality and Anti-Discrimination Ombud (Likestillings- og diskrimineringsombudet, LDO)	*Regulator*: Evaluates technology and discrimination, particularly vulnerability through targeting of advertising
The Consumer Authority (Forbrukertilsynet, FT)	*Regulator*: Provides supervision and guidance in digital marketing, linked to the Marketing Act
The Consumer Council of Norway (Forbrukerrådet, FBR)	*Interest organisation*: Focuses on consumer protection in the advertising industry and platforms, especially mechanisms for (un)fair influence
Norwegian Federation of Organisations of Disabled People (Funksjonshemmedes fellesorganisasjon, FFO)	*Interest organisation*: Interested in the impact of advertising on the disabled and chronically ill
Amnesty International Norway (AIN)	*Interest organisation*: Problematises human rights on global (advertising) platforms and vulnerabilities related to data collection and communication
Norwegian Council for Mental Health (Rådet for psykisk helse, RPH)	*Interest organisation*: Umbrella organisation with a focus on psychological problems, e.g., addiction and representations in advertising

minutes each. Our analysis led to categories that describe the understandings the informants have about behavioural advertising, what interests this form of advertising is linked to, attitudes to the collection of personal data, and the relationship to regulation.

9.3.3 Consumers

In our consumer study, we interviewed (personal interviews and focus groups) 15 individuals from the greater Oslo region. Out of these informants, nine are considered young adults[2] and are included in this analysis. The focus groups and in-depth interviews were conducted during Autumn 2023, recorded, transcribed, and anonymised. The goal was to discuss personalised advertising with consumers recruited on the basis of special/sensitive personal data categories (see Table 9.2) to uncover their experiences with personalised advertisement and advertisement targeting on social media.

Table 9.2 Informant characteristics – consumers

Informant	Age range	Personal data categories	Social media used
Miriam	18–23	Female, person of colour	Instagram, Snapchat, TikTok
Cecilie	24–29	Female, politically active, bisexual	Facebook, Instagram, Snapchat, TikTok
Lene	30–35	Female, parent of a toddler	Facebook, Instagram, Snapchat
Sander	30–35	Male, Parkinson's	Facebook, Instagram, Snapchat
Aram	24–29	Male, person of colour	Facebook, Instagram, Snapchat, TikTok, X/ Twitter
Ingvild	30–35	Female, Coeliac disease, diabetes, bisexual, ADHD	Facebook, Instagram
Tina	30–35	Non-binary, Coeliac disease, lactose intolerance	BeReal, Facebook, Instagram, TikTok
Cathrine	30–35	Female, Parkinson's, mobility impairment	Facebook, Instagram, Snapchat
Lise	30–35	Female, Parkinson's, ADHD	Instagram, Snapchat, LinkedIn

9.4 RESULTS

9.4.1 The Construction of the Advertising Economy, Data and Vulnerability

The Norwegian advertising market is highly impacted by global actors. There are no exact figures for foreign platforms' share of advertising revenue in Norway, but it is assumed to be increasing by the institutional actors interviewed.

> Google and Facebook account for around 40% of the market in Norway. It is challenging for many, and not least for Norwegian media players. (ANFO)

> We compete with … Google and Meta in the advertising market. And, simply put, when they somehow create very advanced data products and use a lot of data to do it, it is simply difficult for us to compete and deliver equally good advertising products to the advertisers. (SBT)

Social and non-editorial media are winners in the so-called 'channel battle' (ANFO, 2023). Norwegian media companies, such as Schibsted quoted above, compete against Google and Meta on two fronts in the digital advertising market. First, the amount and breadth of personal data collection. For instance, Google collects data across devices and through browsers, where approximately '75 percent of all pages have some kind of Google tracker on them', as one of the informants put it. Second, the media companies compete with them in terms of the degree of effectiveness (that is, accuracy and conversion) of advertising campaigns.

Norwegian media companies are struggling to keep up. The global players have greater access to personal data; this is an important competitive advantage when it comes to programmatic advertising. This is largely consistent with the findings in the European Commission's report, which concludes:

> A significant number of digital advertising revenue flows towards large platforms which compete with publishers to sell ads next to the content they host. (European Commission, 2022, p. 10)

For example, Schibsted, as the owner of the largest Norwegian media platform, calls for competition on equal terms.

Interest and non-governmental organisations link their criticism of behaviour-based, programmatic advertising to privacy and, ultimately, to human rights. The most explicit hereto is Amnesty, who in September 2022 demanded the establishment of a national algorithm oversight (an Algorithm Ombudsman) and a 'ban on surveillance-based advertising' (AIN). Supporting arguments are made by advertising agencies with alternative, contextual advertising models. One agency states that privacy

> is a fundamental right we have. And there's a reason it's basic, because the day you need it, you'll be very glad that someone cared for it. And the same applies to freedom of expression. It's not like everyone stands on the barricade for freedom of expression every day. But the day this right disappears, you notice it. (Kobler)

Stronger regulation would be justified from the perspectives of right to privacy, human rights, and freedom of expression. The black box of data for behavioural advertising targeting becomes equated with surveillance and a violation of human rights. Yet, this only applies when it is based on third-party data collection and analysis. As the CEO of Kobler puts it:

> I'm not against behavioural advertising. But if I think about what most people expect is when you go to a website, you have a relationship with that website. And this is called the first-party relationship. Few people have the ability, or the opportunity, to understand that information is sent out to hundreds of third parties. (Kobler)

If the consumers are challenged by a limited understanding of how their data are being collected, advertisers typically believe in the value of detailed data:

> There is a perception that the value of digital advertising is higher when it can be targeted based on detailed knowledge about the individual who is likely to view the ad (usually linked to profiling based on personal data). However, there appears to be limited empirical evidence to support this notion. (European Commission, 2022, p. 136)

As intimate knowledge of the consumers leads to questions of vulnerability, we asked the institutional informants to weigh in on this. Their points of view relate to what we define as positional (personal) and contextual (universal) 'sources' of vulnerability: namely whether vulnerability is linked to specific situations (e.g., where the consumer trades in a market with market power on the seller's side) or whether the consumer is generally vulnerable (Cartwright, 2014). In this context, the starting point for discussing vulnerability will be the factors that the privacy legislation states are sensitive personal data: information about racial or ethnic origin, political opinion, religious or philosophical beliefs, health information, and sexual orientation.

Positional vulnerability has clear links to (the possibility of) being discriminated against and stigmatised. On the one hand, the Equality and Anti-Discrimination Ombud sees discrimination linked to several products and services, which are not offered to certain groups due to specific characteristics. On the other hand, like our informant from the Norwegian Federation of Organisations of Disabled People says, 'it is clear that there are some groups that are more stigmatised than others … mental illness is particularly vulnerable'. Digital worlds seem to 'open completely unusual opportunities to commit abuse against children and young people and other people' (RPH). Problems can be amplified with behaviour-based advertising, where prejudices or assumptions can lead to discrimination:

> We also see that [behaviour-based advertising] can have clear consequences for the right not to be discriminated against. And precisely what kind of categories are used about us, why do we see certain types of content, what kind of content are we not allowed to see. Is there a risk that I do not see certain types of job advertisements because I am a woman, or that I do not see certain types of housing ads because I belong to a certain ethnic minority? Then we are on a track of discrimination, which is also very important to us. And then it opens the door for exploitation of us in vulnerable situations. (AIN)

Where this quote from Amnesty emphasises 'exploitation of us in vulnerable situations', others point out that 'anyone is vulnerable'.

> …most people don't care, because they are not in that situation themselves. But there's a lot about the day-to-day system that can make anyone vulnerable … vulnerable uses … people underestimate their own vulnerability. (Kobler)

Vulnerability comes with clear regulatory implications. Some of these can be internal, 'self-regulating' approaches to segmentation that Schibsted describes:

> Things that have to do with health, sexual orientation and religion, we have defined a long list of various points that go to the type of toxic data points … We have a list of segments we can sell to advertisers, something we have careful audits of. (SBT)

External regulation draws on the General Data Protection Regulation (GDPR) and consent.

> This means that if the letter or wording of the GDPR is to be fulfilled, then there must be an informed consent. This means that you must be able to understand what it is you are giving consent to. It is not possible to understand what consent is given to third-party actors. When we are logged in to Google because you are going to use Gmail … or Google Drive, we are logged in to Google. Then they have the knowledge and insight they need about you anyway. (MBL)

Yet, beyond the reflection on the types of vulnerability or toxicity of data, most governmental and non-governmental organisations had not received complaints from consumers (and citizens) related to actual discrimination in digital advertising. This leads us to focus on the consumer's experiences in digital spaces.

9.4.2 Consumer Perceptions of Agency and Vulnerability

Our consumer informants commonly describe a lack of agency and an ensuing concern regarding their perceived control over their personal data and its use. Agency is particularly linked to actions relating to cookies and website tracking and attempts to exercise control through cookie settings. Here, consumers see their efforts producing limited rewards. Cathrine, for instance, says she has tried to personalise cookies:

> And I only click essential [cookies] or deny, but while you're in there, they've snuck in that you think you're clicking confirm on it – on 'only essential cookies', then instead it says 'accept all' or something … I haven't thought I accepted that much, but I probably accepted to sell my whole soul. (Cathrine)

Ingvild also says that despite having intentions of maintaining stricter data privacy, she finds it challenging to block out all possible sources of tracking. Therefore, she feels resigned.

> Say, for instance, I use DuckDuckGo, and have Proton mail instead, but then I use Google Maps because I want those tips about the restaurants that are relevant for me … And then I have already sold out everything, I feel. So, it feels like, either I have to remove myself completely and just try to not be digital, almost not at all, or I have to accept that it is just how it is. (Ingvild)

The only glimpse of agency that Ingvild's account points to is resigning either all digital presence or any illusion of being able to manage the movement of data. When reflecting upon cookies, Lise says that she has closed her eyes and pressed yes.

> I don't want to, but I can't be bothered to personalise those f*****g cookies. I accept. Accept, accept.

By saying this, Lise highlights that she struggles to utilise the tools for agency that are available to her. Aram describes a similar lack of results from his active effort in deleting cookies.

> I went and deleted a good deal of things, just like 'why not?'. Then [personalised advertisements] disappeared, advertisements became general, but it took max six months then they were back again. Then I tried again. There was [personalised] advertisement from week two … because they built up my profile again. (Aram)

While some of the informants may differ slightly in the way they attempt to secure their data privacy, as shown above, they experience limited agency. The question then becomes: what are the consequences? The limited agency these informants experience is linked to feelings of vulnerability. Our informants have several categories of potential vulnerability through ethnicity, sexual preference, or chronic health condition. Our informants are aware of their potentially amplified vulnerability and share a concern for their data privacy – who collects and stores their personal data and how it is being utilised.

While certain categories of sensitive data may be considered 'toxic' and as such should not be used for profiling and targeting (as our industry informants elaborate), several of the informants are more concerned about the potential misuse of aggregated user data rather than singular 'toxic' data. More specifically, the informants describe potential scenarios in which they personally are at risk due to the data associated with them:

> It has become very popular online to have different gender categories. I'm non-binary, but it is very popular now to have it [as non-binary gender category] as a way to seem more progressive and inclusive. But I don't like to have it as a registered factor in any way. I don't think we should have gender in any documents, either, because like, if there now should come a government who want to find out who is trans, who is this, who is that, then you have page upon page of registers – and that has happened previously… (Tina)

In the larger context of the interview, Tina is drawing a historical parallel with data collection, data registries, and societies in which citizens have been subject to systematic discrimination and persecution based on registered personal data: Tina suggests that the personal data consumers reveal, and that various organisation collect and have access to, may be (mis)used in a similar manner. Others, such as Ingvild, put this idea more explicitly (and with a dark, dystopian twist) by using themselves as an example:

> It is more the sum of it [personal data], I think. The sum can, like, to overstate a bit, and what is hopefully just pure fantasy, but if it in the future becomes like … That we don't have space for everyone on this earth, for instance, and that we only must use those most capable, then I would quickly belong to a category that isn't that capable because I have diabetes, ADHD and like… (Ingvild)

Another feeling of vulnerability concerns the informants as individuals in the present. Aram represents this position. He is concerned about the societal implications of data collection as part of programmatic, behavioural targeting of advertising. He points out how personalised advertising encroaches on personal identity:

> It is the collection of all data points which is problematic to me. They collect data from everywhere, really. Both from location, interests, what you've searched for and what you haven't searched for. There are so many datapoints they try to use.

> That is what annoys me, you feel that there is no place I can be without having my story tried and summarised. (Aram)

A specific instance of Aram having his story 'tried and summarised' is his experience with university advertisements. He does not make it clear whether or not he has clicked on said advertisements. Nevertheless, because he does not know what kind of master's programme he wishes to opt for, he feels the university advertisements he receives are poorly targeted and therefore irritating:

> I barely know that I'm going to do a master's, and then they like ... They just throw everything. Had I clicked on some of them, or on two of them, then that would maybe have given them more information. But the fact that I haven't clicked on the ad makes it so that they try to send as many [advertisements] as possible, to see what happens. Throw everything my way and see what sticks... (Aram)

While he finds it problematic and annoying that 'they' attempt to identify and show him personalised advertisements, he endeavours to hinder data collection by resetting cookies. This somewhat contradictory behaviour, in which both data security and better targeted advertisements are desired, is in contrast with the sense of resignation other informants experience. Lise describes this as feeling that 'you completely expose yourself' regarding the personal data she has revealed to 'them'. Ingvild explores this point in more detail:

> It irritates me that they can know everything about me, really. But I understand that we are in a way f****d already, a long time ago. And if I'm f****d, then there are a lot of others who are much more f****d than me, long before me. I feel that I just cannot be bothered. I know, kind of, what I'm doing, as long as I don't do anything criminal that someone can catch me doing, then I feel that I could not care less.

By saying this, Ingvild describes a complete resignation of agency.

All in all, the concern for societal and individual consequences following from the worry over potential misuse of personal data showcases that the effects of limited agency may lead to experienced vulnerabilities. Exceptions to this are Lene and Miriam, both of whom perceive data collection as less problematic than the rest of our informants. As showcased below, this has implications for Lene and Miriam's sense of agency, and subsequent vulnerability. While she does link data collection to the advertisements she sees, Miriam does not see the collection of data as problematic:

> [discussing the informant's consumption of cosmetics]
>
> Miriam: I search a lot for those kinds of products, face spray and such, and then it appears on almost all platforms. I haven't thought much about

that it may be connected to what I look up on Google, but it is very nice to see that it's appeared. I've just gotten more options.

Interviewer: It's mostly when you've looked things up you experience personalised advertisement, which concerns exactly what you've looked up?

Miriam: Yes, I think that's nice.

As established above, Miriam responds positively to how personal advertisements utilise her digital actions, a finding consistent with those of Ruckenstein and Granroth (2019). Given her positive experiences with and unconcerned attitude towards data collection, she experiences agency in feeding the algorithms content for which she wants to see advertisements; effectively, this does not result in her feeling vulnerable. When asked if the kind and colour of makeup she sees advertisements for are connected to her skin tone or her country of origin, deduced through web activity, Miriam responds: 'Yes. Actually, that's fine. If they've checked my phone, seen where I come from rather than looking at my history, that's good. That's fine.' At the end of the interview and after establishing Miriam's unconcerned attitude towards personalised advertising, the interviewer wished to test how she would respond to the advertising industry's legal barriers. It is only when explained that deducing ethnic and racial identities from judicially non-sensitive data (such as location, interests, and web activity) and using this in marketing is illegal that Miriam reacts negatively by saying:

> What you [interviewer] are saying now is that they collect data to find out what race [you are], and that is wrong? And that they actually can't advertise that way? That's discrimination. It kind of is discrimination, that they kind of separate you, because you're black. 'No, that's because you're white'. That's not good. (Miriam)

Up until this point in the interview, Miriam showed little to no apprehension in giving over personal data. Her unconcerned attitude towards data collection is based on a limited understanding of the legal limits of personalised advertising and data collection. Therefore, she does not experience the vulnerability that the other informants do. Miriam's attitude also illustrates that she feels her personal data are used well in personalising advertisements, and that she therefore experiences agency in 'forming' the algorithms and the advertisements she sees. In this sense, the theoretical dynamic between agency and vulnerability applicable to the other informants is not applicable to Miriam.

Another example of this is Lene, a recent mother, who sees data collection as slightly problematic and experiences partial agency and vulnerability. She feels that she influences what advertisements she receives, and appreciates that by saying:

> If I look up a pram, then I know that I'll get loads of baby things in my feed, and probably will for a year or two more [laughs] – and I think that's completely fine, because it is relevant. (Lene)

At the same time, Lene sees the potential downsides of data collection as well: 'I think mainly, loads of things. Sexual orientation and for instance sexual preferences, or diseases, or something you don't actively write on Facebook.' Later, Lene says: 'What if you look up loads of [things] that you don't want others to know about, that you look up?'. In saying this, Lene sees the potential for secrets one 'shares' with the internet in confidentiality being unwillingly spread to others, and thusly acknowledges this as a source of vulnerability. However, she also states that she is not worried about it, as 'I think that it cannot be anything less interesting than to keep track of what I'm doing. If they keep track of several millions of people, then there is nobody that notices what I'm doing'. All in all, while Lene does not perceive data collection as unproblematic to the same degree as Miriam, she does perceive data collection as slightly problematic. Lene seems to reduce the experience of limited agency and vulnerability as opposed to most informants.

9.5 DISCUSSION

Personalised, behaviour-based, data-driven programmatic advertising is a hugely complex issue that impacts consumers in their everyday lives while working on a global level. This chapter examines the complexity in personalised advertising and behavioural targeting, and how consumers consider themselves 'seen' by the algorithm and react to it while considering their vulnerabilities through sensitive data and their potential amplification.

While some argue that the value of individual data points is diminishing (Lovink, 2019) and others claim that investments made in 'more effective' digital, behaviour-based, programmatic advertising are misguided (ANA, 2023), our analysis highlights a juncture between the ambivalence of consumers regarding their personal data and personalised advertising, and the conflicting industry narratives around the need and effectiveness of behaviour-based advertising. Here, vulnerability is discussed in both positional and contextual terms, referring to the characteristics of the individual consumer and the situation from which the data emerges, respectively.

From our data, the link between agency and vulnerability is grounded in the observation that when young consumers experience a lack of control of their own personal data, they also experience a vulnerability at the hands of those who may potentially exploit their data. By describing the ubiquitous nature of data collection – including gender, skin tone, and medical diagnoses – the informants point to the intersectional nature of their experienced vulnerabilities. Their vulnerabilities are not only caused by a lack of agency regarding one data category but by a combination of the categories. This effect may have a circular impact – a feedback loop – where feeling vulnerable diminishes one's sense of agency and one's willingness to attempt to utilise the tools of agency available. In the case of young adult consumers, the lack of agency is directly related to the agency of those who possess their personal data.

In our analysis of the consumer data, we find stances towards personalised, programmatic advertising also found in other contributions (Büchi et al., 2023; Lomborg & Kapsch, 2020; Ruckenstein & Granroth, 2019; Sörum & Fuentes, 2023). These studies point to a positive stance towards personalising ('being seen' by the algorithm); an ambivalent stance towards inaccurate profiling ('being seen' incorrectly), and a negative position towards personalising that borders on a data breach. We find two understandings of data breach – a resigned and an oppositional stance – leading to experiences of vulnerability. While this categorisation is less than novel, part of our contribution is explicitly linking consumers' experiences of lacking agency to their experiences of vulnerability.

9.5.1 Practical Implications

Our research also points to gaps in practical approaches to informal and formal regulation. With the current regulatory framework, greater demands will be placed on consumers' own ability to protect privacy through declarations of consent. Our analysis of user experiences raises questions about whether consumers are truly equipped to safeguard their interests adequately. The complexity of data collection with (and without) consent places significant demands on the consumer's knowledge. Often, additional information and control mechanisms only mean more control work for consumers, while they desire less work and responsibility. We find that the complexity in the business practices of personalised advertising does little to facilitate a rational balance between cost and benefit, or to assess information that should make visible the connection between data and advertising. This means that the risk of sharing data is systematically underestimated. We also find that personalised advertising, intentionally or unintentionally, uses methods such as consent exhaustion, which lead to powerlessness and resignation.

9.5.2 Avenues for Future Research

While our findings illustrate a nuanced understanding of vulnerability, adding to previous (Nordic) research discussing the understanding of consumer profiling and targeted digital advertising (e.g., Lomborg & Kapsch, 2020;

Ruckenstein & Granroth, 2019), and more generally illustrate practical implications applicable to most affluent post-industrialised societies, there are limitations to the study. These pertain to the limited number of informants in this study, their perception (and reflection) of their own vulnerability, the valuation of their data, and the externalisation of relationality in the process of personalisation. Our informants tend to talk about 'them', pointing to an understanding of privacy in digital media as less than personal, and specifically not interpersonal. We also question whether the regulatory burden limits consumers' ability to self-regulate. Put differently, a low degree of consumer agency equals a high degree of agency for the algorithms, the advertisers, and ad tech actors who save and utilise consumers' personal data. This power dynamic is at the centre of the vulnerabilities the young consumers experience and should be the focus of market and consumer privacy regulating interventions as well as future research.

NOTES

1. The data in this study were collected in connection with a research project entitled 'Why am I seeing this ad?' which was commissioned by the Norwegian Ministry of Children and Families in 2023. The project's aim was to create a better understanding of which data are used in personally targeted marketing in digital channels, how these may be linked to users' life situations, and whether the use of special category personal data could be shown. The commissioned work connects to the digital priorities in the Norwegian context (see section 9.3.1 Research Context).
2. We extend the category of young adults to individuals up to 35 years of age; the informants within this category range from 18 to 35 years old. Due to reasons of anonymity, we do not state the exact age of informants but give these in three age range brackets. Pseudonyms have been used.

REFERENCES

ANA. (2023). *ANA programmatic media supply chain transparency study: Complete report*. Association of National Advertisers. https://www.ana.net/miccontent/show/id/rr-2023-12-ana-programmatic-media-supply-chain-transparency-study.

Andrejevic, M. (2014). Big data, big questions. The big data divide. *International Journal of Communication*, *8*, 1673–1689.

ANFO. (2023). *Den store annonsørrapporten*. ANFO Annonsørforeningen.

Barnes, S. B. (2006). *A privacy paradox: Social networking in the United States*. First Monday. https://doi.org/10.5210/fm.v11i9.1394.

Batat, W. (2010). *Understanding the dimensions of young consumer vulnerability in the web 2.0 society*, *250*. https://shs.hal.science/halshs-00527884.

Büchi, M., Fosch-Villaronga, E., Lutz, C., Tamò-Larrieux, A., & Velidi, S. (2023). Making sense of algorithmic profiling: User perceptions on Facebook. *Information,*

Communication & Society, *26*(4), 809–825. https://www.tandfonline.com/doi/abs/10.1080/1369118X.2021.1989011.

Carter, D., & Eger, E. K. (2021). Visibility and vulnerability in online marketing practices. *Journal of Cultural Economy*, *14*(4), 373–387. https://doi.org/10.1080/17530350.2021.1879212.

Cartwright, P. (2014). Understanding and protecting vulnerable financial consumers. *Journal of Consumer Policy*, *38*(2), 119–138. https://doi.org/10.1007/s10603-014-9278-9.

Colicev, A., Malshe, A., Pauwels, K., & O'Connor, P. (2018). Improving consumer mindset metrics and shareholder value through social media: The different roles of owned and earned media. *Journal of Marketing*, *82*(1), 37–56. https://doi.org/10.1509/jm.16.0055.

Datatilsynet. (2015). *Det store datakappløpet. Rapport om hvordan kommersiell bruk av personopplysninger utfordrer personvernet, november 2015*. Datatilsynet. https://www.datatilsynet.no/globalassets/global/dokumenter-pdfer-skjema-ol/rettigheter-og-plikter/rapporter/kommersialisering-norsk-endelig.pdf.

Dentsu. (2023). *Ad spend 2023: Dentsu global ad spend forecasts 2023*. https://insight.dentsu.com/ad-spend-may-2023/.

Draper, N. A., & Turow, J. (2019). The corporate cultivation of digital resignation. *New Media & Society*, *21*(8), 1824–1839. https://doi.org/10.1177/1461444819833331.

European Commission. (2022). *Behavioural study on unfair commercial practices in the digital environment: Dark patterns and manipulative personalisation: Final report*. Publications Office of the European Union. https://data.europa.eu/doi/10.2838/859030.

Farahat, A., & Bailey, M. C. (2012). How effective is targeted advertising? *Proceedings of the 21st International Conference on World Wide Web*, 111–120. https://doi.org/10.1145/2187836.2187852.

Forbrukerrådet. (2020). *Out of control: How consumers are exploited by the online advertising industry*. Forbrukerrådet. https://storage02.forbrukerradet.no/media/2020/01/2020-01-14-out-of-control-final-version.pdf.

Forbrukerrådet. (2021). *Time to ban surveillance-based advertising: A case against commercial surveillance online*. Forbrukerrådet. https://storage02.forbrukerradet.no/media/2021/06/20210622-final-report-time-to-ban-surveillance-based-advertising.pdf.

Fuchs, C. (2017). *Social media: A critical introduction* (2nd ed.). Sage. https://fuchsc.net/books/2nd-edition-social-media-a-critical-introduction/.

Hackley, C., & Hackley, R. A. (2021). *Advertising and promotion* (5th ed.). Sage.

ICO. (2019). *Update report into adtech and real time bidding*. Information Commissioner's Office.

Jenkins, H. (2006). *Convergence culture: Where old and new media collide*. NYU Press. https://www.jstor.org/stable/j.ctt9qffwr.

Lee, H., & Cho, C.-H. (2020). Digital advertising: Present and future prospects. *International Journal of Advertising*, *39*(3), 332–341. https://doi.org/10.1080/02650487.2019.1642015.

Lomborg, S., & Kapsch, P. H. (2020). Decoding algorithms. *Media, Culture & Society*, *42*(5), 745–761. https://doi.org/10.1177/0163443719855301.

Lovink, G. (2019). *Sad by design*. Pluto Press. https://red-thread.org/wp-content/uploads/2020/05/Lovink_en.pdf.

Lupton, D. (2021). 'Not the real me': Social imaginaries of personal data profiling. *Cultural Sociology*, *15*(1), 3–21. https://doi.org/10.1177/1749975520939779.

Lynch, L. (2018). Native advertising: Advertorial disruption in the 21st-century news feed. In *Native advertising: Advertorial disruption in the 21st century news feed*. https://doi.org/10.4324/9781315175119.

Medienorge. (2024). *Bruk av sosiale medier en gjennomsnittsdag*. https://medienorge.uib.no/statistikk/medium/ikt/412.

Milyaeva, S., & Neyland, D. (2016). Market innovation as framing, productive friction and bricolage: An exploration of the personal data market. *Journal of Cultural Economy*, *9*(3), 229–244. https://doi.org/10.1080/17530350.2015.1135473.

O'Day, E. B., & Heimberg, R. G. (2021). Social media use, social anxiety, and loneliness: A systematic review. *Computers in Human Behavior Reports*, *3*, 100070. https://doi.org/10.1016/j.chbr.2021.100070.

OECD. (2023). Consumer vulnerability in the digital age. *OECD Digital Economy Papers*. https://www.oecd.org/publications/consumer-vulnerability-in-the-digital-age-4d013cc5-en.htm.

Rounsefell, K., Gibson, S., McLean, S., Blair, M., Molenaar, A., Brennan, L., Truby, H., & McCaffrey, T. A. (2020). Social media, body image and food choices in healthy young adults: A mixed methods systematic review. *Nutrition & Dietetics*, *77*(1), 19–40. https://doi.org/10.1111/1747–0080.12581.

Ruckenstein, M., & Granroth, J. (2019). Algorithms, advertising and the intimacy of surveillance. *Journal of Cultural Economy*, *13*, 1–13. https://doi.org/10.1080/17530350.2019.1574866.

Samuel, A., White, G. R. T., Thomas, R., & Jones, P. (2021). Programmatic advertising: An exegesis of consumer concerns. *Computers in Human Behavior*, *116*, 106657. https://doi.org/10.1016/j.chb.2020.106657.

Schiro, E. C. (2023). *Norsk Mediebarometer 2024*. Statistics Norway.

Slettemeås, D., & Storm-Mathiesen, A. (2021). Digitalisert forbruk. In G. Schjelderup (Ed.), *Forbrukersosiologi. Bærekraft, digitalisering, identitet og makt* (pp. 447–474). Fagbokforlaget.

Smit, E. G., Van Noort, G., & Voorveld, H. A. M. (2014). Understanding online behavioural advertising: User knowledge, privacy concerns and online coping behaviour in Europe. *Computers in Human Behavior*, *32*, 15–22. https://doi.org/10.1016/j.chb.2013.11.008.

Sörum, N., & Fuentes, C. (2023). How sociotechnical imaginaries shape consumers' experiences of and responses to commercial data collection practices. *Consumption Markets & Culture*, *26*(1), 24–46. https://www.tandfonline.com/doi/abs/10.1080/10253866.2022.2124977.

Spiekermann, S., & Christl, W. (2016). *Networks of control – A report on corporate surveillance, digital tracking*. Facultas.

Statistics Norway. (2023, September 22). *Bruk av IKT i husholdningene*. SSB. https://www.ssb.no/teknologi-og-innovasjon/informasjons-og-kommunikasjonsteknologi-ikt/statistikk/bruk-av-ikt-i-husholdningene.

West, S. M. (2019). Data capitalism: Redefining the logics of surveillance and privacy. *Business & Society*, *58*(1), 20–41. https://doi.org/10.1177/0007650317718185.

Zuboff, S. (2015). Big other: Surveillance capitalism and the prospects of an information civilization. *Journal of Information Technology*, *30*(1), 75–89. https://doi.org/10.1057/jit.2015.5.

10. Factors preventing young people from protecting their commercial privacy in digital environments

Sonali Srivastava

10.1 INTRODUCTION

Young people have a significant presence in digital environments (Stoilova et al., 2020). According to a blog post by Children Online Research and Evidence (CORE), digital environment is an ever-evolving term that encompasses information and communication technologies, such as digital networks, various services and applications, and connected devices, among others (see, for example, Stoilova & Livingstone, 2022). Young people's increased presence in digital environments provides them with various opportunities to build peer and other interpersonal relationships, develop and express their identities, access educational content and entertainment opportunities, and engage in civic participation (Milkaite & Lievens, 2019; Hollis et al., 2020). However, contemporary digital environments are characterized by underlying covert data-gathering practices that form the basis of commercial profiling and targeting (Stoilova et al., 2019a, 2020). Commercial privacy is defined as "how my personal data is harvested and used for business and marketing purposes" (Livingstone et al., 2019b, p. 4). This chapter aims to explore the factors that prevent young people from protecting their commercial privacy in digital environments.

Commercial data-gathering practices in digital environments leave young people particularly vulnerable for three main reasons. First, such practices not only endanger young people's data privacy and security (Livingstone et al., 2019a) but also their right to privacy (Lievens, 2018). Second, profiling young people based on their previous online activities and subsequently targeting them with content and advertisements on the basis of their data profiles can be particularly deleterious as it can limit young people's access to multiple choices and perspectives at a stage in their lives when they develop their identities and make crucial life choices (Milkaite & Lievens, 2019). Finally, these practices enhance young people's susceptibility to commercial persuasion (Van der Hof et al., 2020). However, young people's commercial privacy in digital environments is under-researched (Stoilova et al., 2019b).

Previous research has shown that young people's inability to conceptualize commercial actors as harmful (Keen, 2020), faulty theories regarding how online user data-gathering processes work (Holvoet et al., 2021) and the powerlessness they experience due to lengthy privacy terms (Stoilova et al., 2020) largely prevent them from protecting their commercial privacy in digital environments. However, previous research on the factors preventing young people from protecting their online commercial privacy does not comprehensively discuss how dark patterns in digital environments make privacy protection inconvenient or data sharing convenient for young users. Dark patterns are "interface design choices that benefit an online service by coercing, steering or deceiving users into making decisions that, if fully informed and capable of selecting alternatives, they might not make" (Mathur et al., 2019, p. 2).

This chapter contributes to research on the factors that prevent young people from protecting their commercial privacy in digital environments. It does so by highlighting how the use of dark patterns in digital environments makes privacy protection inconvenient or giving away data convenient for young people. This chapter contributes to the section on challenges for young people in digital environments. It does so by highlighting that besides young people's knowledge gaps, the design of digital spaces that make privacy protection inconvenient prevents young people from protecting their online commercial privacy and poses challenges for young people in digital environments. The chapter uses data from eight focus group discussions (FGD) (*N*=38) conducted in Finland's capital area between December 2020 and May 2021 and draws primarily from the results of a published study (in which the author of this chapter is the first author) exploring children and young people's (13–16 years) privacy negotiation practices in the digital commercial context (see Srivastava et al., 2023a).

10.2 LITERATURE REVIEW

10.2.1 Privacy in the Digital Commercial Context and Dark Patterns

User data collection in the digital commercial context is characterized by a certain opacity (Stoilova et al., 2020). According to Van der Hof (2016), this opacity is largely because data for commercial profiling and ensuing targeting is not just gathered when it is explicitly provided by users. Additionally, users' online activities (such as their search history, the videos they watch, what they like, and so on) are tracked through the use of various tracking technologies such as cookies (Van der Hof, 2016). Acquisti et al. (2020) note that tracking technologies are also constantly evolving and getting more advanced. For instance, when users tried to avoid "cross-site tracking", new cookies (e.g., flash cookies) were introduced (Acquisti et al., 2020, p. 746). The opacity of online user data collection for commercial purposes contributes to information asymmetries because users are not fully aware of how their data are collected, shared, and used (Acquisti et al., 2020; Waldman, 2018). Therefore, while increasing awareness of online data collection practices and data protection methods is essential for effective privacy management (Stoilova et al., 2020), the inherent information asymmetries that underpin digital commercial data-gathering practices should also be addressed through regulatory efforts (Acquisti et al., 2020; Stoilova et al., 2019b).

Regulations have attempted to minimize information asymmetries. The General Data Protection Regulation (GDPR) in the EU, for instance, requires that privacy notifications be provided, and users' consent be sought before gathering their data (Livingstone, 2018). However, privacy notifications do not offer users much real choice as they are often lengthy and expressed in complex language, which makes them unintelligible for ordinary users (Van der Hof, 2016; Waldman, 2018). Since clear guidelines on how to seek consent or design privacy notifications (also called cookie notifications) are largely absent, dark patterns are common (Berens et al., 2024). The term *dark patterns* was introduced in 2010 by Harry Brignull (Mathur et al., 2021). Dark patterns in cookie notifications make users select an option that leads them to give more data than they would prefer to give (Berens et al., 2024). Waldman (2020) notes that the design of digital environments can constrain users' ability to protect their privacy. Dark patterns are highly instrumental in making users part with more data than they would provide if they were fully or clearly informed of the consequences of doing so (Waldman, 2020). One of the attributes of dark patterns is "information hiding", which essentially involves obscuring or delaying users' access to essential information (Mathur et al., 2021, p. 8).

There have been calls for making digital environments more conducive to privacy protection (Waldman, 2020). For instance, the need for more precise regulatory guidelines on how to design privacy notices to make consent meaningful (Utz et al., 2019) and increasing designers' responsibility and ethical conduct to avoid dark patterns while designing digital environments (Gray et al., 2020) has been underscored. It has been stressed that special attention should be paid to young people's privacy needs in digital environments (Livingstone et al., 2019a; Stoilova et al., 2020). Furthermore, it has been emphasized that corporations that provide media services and content for children and young people should consider it their corporate social responsibility to consult young people and prioritize their interests and needs, including their need for online safety (Lievens, 2018).

10.2.2 Research on Factors that Prevent Young People from Protecting their Commercial Privacy in Digital Environments

Previous research has identified some factors that prevent young people from protecting their commercial privacy in digital environments. Keen (2020), drawing from interviews with participants aged 12–16 years, explores whether young people are aware of being monitored online and the harms they perceive from such practices. Keen notes that the biggest factor preventing young people from protecting their privacy is their inability to conceptualize corporations as harmful. Young people in Keen's study viewed corporations as impersonal and non-judgmental entities. Moreover, the participants thought that the data corporations collected was unlikely to seep into interpersonal contexts and cause any reputational harm, thus leading them to view data gathering by commercial entities as innocuous. Unless a conceptual shift in young people's ways of thinking about privacy harms from commercial actors takes place, privacy self-management is likely to remain ineffective (Keen, 2020). However, Keen (2020) does not specifically explore young people's commercial privacy management practices. Holvoet et al. (2021) do so, and their study shows that young people's reliance on folk theories or inaccurate understandings of reality (for instance, thinking that commercial data gathering is inevitable, and all privacy policies are the same) impedes effective online commercial privacy protection. They note that most of their participants just accepted cookies without paying attention to them because they were in a rush to move on and pursue their online activities. Their participants did not consider editing privacy terms as a necessary means of protecting their online privacy. Raising young people's awareness is important, along with displaying icons that warn them of commercial data use (Holvoet et al., 2021). Therefore, it might be safe to conclude that while discussing the factors that prevent young people from managing their online commercial privacy, both of the aforementioned studies focus on issues that are primarily related to young people's knowledge gaps.

Stoilova et al. (2020) have researched young people's (11–16 years) commercial privacy management in digital environments. They highlight knowledge gaps and issues with conceptualization. They note, for instance, that young people conceptualize commercial privacy in the same way as they perceive interpersonal privacy (which is online privacy in relation to known people such as peers, parents, teachers, and so on), in that they tend to view commercial data collectors as individuals who would act responsibly (Stoilova et al., 2020). They observe that only a few young people seem to have an awareness of how cookies assist in the commercial data-gathering process. Stoilova et al. posit that young people's knowledge gaps need to be addressed through media literacy education, but the state and businesses must also play a focal role in providing a transparent digital environment that offers young users real choices for privacy protection. Regarding the role of businesses, they pinpoint that incomprehensible privacy notifications, among others, contribute towards making digital environments disempowering for young people (Stoilova et al., 2020), thus alluding to how contemporary digital environments do not support commercial privacy protection. Like Stoilova et al. (2020), Holvoet et al. (2021) also make suggestions for making digital spaces more conducive to privacy protection. They note that some warning mechanisms or symbols could be used to indicate to young users that their data are being collected. However, Holvoet et al. (2021) mainly focus on highlighting young people's lack of knowledge, which they consider to be the most significant deterrent to young people's commercial privacy protection in digital environments.

To sum up, according to previous research, the factors that prevent young people from protecting their commercial privacy in digital environments fall mainly under the category of young people's knowledge gaps and how digital spaces may not enable privacy protection. However, previous research on the factors preventing young people from protecting their online commercial privacy does not comprehensively discuss how dark patterns in digital environments make privacy protection inconvenient or data sharing convenient for young users. This chapter addresses this gap in the existing research and poses the following research question:

> RQ: What are the factors that prevent young people from protecting their commercial privacy in digital environments?

10.3 DATA AND METHODS

While the chapter draws primarily from the results of a published study on young people's privacy negotiation practices in the digital commercial context (see Srivastava et al., 2023a), it also further elaborates on some points mentioned by the participants during the focus group discussions (FGDs). Therefore, this chapter is distinct from the published study because it discusses factors that prevent young people from protecting their commercial privacy in digital environments. The study uses data from eight FGDs (*N*=38) conducted in schools located in Finland's capital region.

The FGDs were conducted between December 2020 and May 2021. The participants (13–16 years) were recruited by their teachers. This age group was specifically selected for the study as the GDPR stipulates that member nations can fix a digital age of consent at a minimum of 13 years and a maximum of 16 years, thus presuming that young people in this age bracket can effectively manage consent terms (Livingstone, 2018; Keen, 2020). In Finland, the age of digital consent is 13 years (Livingstone, 2018). FGDs were specifically chosen as the method of data collection because they are akin to peer group situations, which are more familiar to young people (Horner, 2000). Moreover, FGDs help in exploring participants' perspectives through an interactive process (Gibson, 2012).

Three schools participated in the study. One of the schools was bilingual (English and Finnish).[1] Informed consent was gathered before the focus group discussions.[2] A small group size of 4–6 participants is considered ideal because peer group interactions usually take place in such group sizes (Eder & Fingerson, 2002). Therefore, each mixed-gender focus group consisted of 4–6 participants. The composition of each group is provided in Table 10.1.

The topics discussed mainly centered on common scenarios in digital environments. For instance, screenshots of everyday situations – such as websites bearing a privacy notice, and commonly available registration options for applications and websites – were shown, and how the participants usually acted in these situations was discussed. Screenshots of targeted advertisements and product suggestions during web surfing were also presented. There were discussions about why specific advertisements appeared, and if the participants had ever tried to make these advertisements incompatible with their preferences or if they had attempted to limit the information they provided to the internet. Special care was taken to ensure that the language used by the moderator was simple and not highly technical. The researcher's influence on participants' responses was minimized by ensuring that the moderator avoided words such as privacy, risk, and harm.

All the FGDs were audio recorded, and each FGD lasted an average of 50 minutes. The audio-recorded FGD data were transcribed and anonymized. FGDs in Finnish were translated into English. The FGD data were analyzed by using thematic analysis (Braun & Clarke, 2006). The method entails

Table 10.1 *Summary of group composition*

Group	School	Boys	Girls	Age (years)
1	Bilingual	1	4	13–14
2	Bilingual	2	3	14–15
3	Bilingual	2	2	15–16
4	Finnish	0	4	15–16
5	Finnish	3	1	15–16
6	Finnish	2	3	15–16
7	Finnish	2	3	14–15
8	Finnish	2	4	15–16

Note: Reproduced from Srivastava et al. (2023a).

familiarizing oneself with the data by carefully reading the transcripts multiple times, coding the data, paying attention to recurring themes across the data, revising these themes after careful re-reading, finalizing the themes, and presenting the report (Braun & Clarke, 2006). The process of coding and identifying themes was performed manually. The results of how children negotiate their privacy in the digital commercial context (Srivastava et al., 2023a) reflected two overarching themes, which could explain the factors that prevent the participants from protecting their commercial privacy in digital environments. Therefore, while describing the findings, there is mention of results from Srivastava et al. (2023a). However, while describing these themes, a further discussion of participants' views is also presented, which helps in better understanding the factors that prevented the participants from protecting their online commercial privacy.

10.4 FINDINGS: FACTORS PREVENTING YOUNG PEOPLE FROM PROTECTING THEIR COMMERCIAL PRIVACY IN DIGITAL ENVIRONMENTS

10.4.1 Knowledge Gaps

Participants' accounts indicated that most young people understood that their previous online activities and location information, among others, were used for profiling and subsequently targeting them with advertisements (Srivastava et al., 2023b). However, the knowledge of cookies was rather non-uniform among the participants. It sometimes varied even among participants belonging to the same focus group. While a few participants explained what cookies did, some others seemed unclear about this. They even acknowledged their lack of awareness, which they thought could hamper their perception of harm from online commercial data gathering.

> [...], maybe my knowledge of the cookies, of the terms, is less than Eva's [another participant in the group] which would mean she is more cautious about it, and I just press it. (Tapio, 15)

Furthermore, knowledge of different types of cookies could also be increased, especially because some participants reported "accepting all cookies" on big companies' websites unhesitatingly because they trusted those companies, which reflects a lack of awareness that even big or known corporations may use third-party cookies (Srivastava et al., 2023a).

Some participants also mentioned accepting cookies to quickly proceed with their web surfing or any other online activity they wanted to engage in

(Srivastava et al., 2023a). This reflects young people's lack of understanding of how editing privacy terms may help them protect their privacy.

Second, knowledge of privacy protection mechanisms could also be increased, as few participants reported adopting measures to control the data they gave away for commercial profiling by, for example, using a virtual private network (VPN) and deleting search history on Google etc. (Srivastava et al., 2023a).

While discussing their choice of registering options on websites, social login was also discussed. Social login is a mechanism that allows users to access an application or a website by using their existing accounts for a platform such as Facebook (Krämer et al., 2019). The participants mentioned that when they were presented with the option of registering on websites by using Google or Facebook or creating their own accounts, they frequently used Google or Facebook to register instead of creating their own accounts (Srivastava et al., 2023a). Using Google or Facebook leads to the transfer of one's information from these platforms to the website in question (Krämer et al., 2019) and may also inform these platforms about a user's activities on these websites (Krämer et al., 2019; Karegar et al., 2018). Hence, social logins can expose users' data (Karegar et al., 2018). Therefore, not pausing to think before using these sign-in options or not treating them suspiciously reflects young people's lack of knowledge regarding how their online data moves (Srivastava et al., 2023a).

10.4.2 Privacy Protection Made Inconvenient

While discussing their response to cookie notifications, various participants talked about how the lengthy and incomprehensible privacy notifications discouraged them from editing the terms (Srivastava et al., 2023a). Some participants thought that these notifications were deliberately made difficult so that accepting cookies becomes the convenient option.

> I feel like they have deliberately made it complicated to understand the cookies so that people just accept them [...]. (Miri, 15)

While discussing social logins, the participants mentioned that they chose the option of using Google or Facebook because it was more convenient (Srivastava et al., 2023a). These options were considered convenient by the participants because the passwords could be easily filled in and the cumbersome process of making new accounts could be avoided.

> Yeah, like if you use Google then normally it comes up with your passwords already, so you don't need to fill it in. (Mikko, 14)

Another participant in a separate group echoed a similar point.

> Yeah, the Google one is much easier than filling it out individually. (Tiia, 15)

Therefore, the options that were convenient to use were also those that led to data sharing, thus making privacy protection inconvenient.

10.5 DISCUSSION

This chapter explored the factors that prevent young people from protecting their commercial privacy in digital environments. A thematic analysis of eight FGDs (*N*=38) with participants (13–16 years) residing in the Finnish capital region suggests that knowledge gaps and how privacy protection is made inconvenient in digital spaces are the two factors that prevent young people from protecting their commercial privacy in digital environments.

The findings on knowledge gaps are consistent with previous research, which has highlighted young people's knowledge gaps as a deterrent to privacy protection (Keen, 2020; Holvoet et al., 2021; Stoilova et al., 2020). Previous research has also paid attention to the powerlessness caused by lengthy privacy terms (Stoilova et al., 2020), underscoring the need to attend to the design of digital environments. However, scant attention has been paid to how dark patterns in digital environments make privacy protection inconvenient or data sharing convenient for young users. This chapter focuses on this issue. First, it highlights how lengthy and incomprehensible cookie notifications do not just create powerlessness, as Stoilova et al. (2020) note. Some participants also saw cookie notifications as being *deliberately* made long and unintelligible so that accepting cookies would become the convenient choice. Dark patterns in cookie notifications push users to choose the option that would benefit corporations (Berens et al., 2024). Since the participants found these notifications to be made long and cryptic on purpose, so that accepting cookies becomes the easier option, these cookie notifications are an example of dark patterns in digital environments. This dark pattern makes privacy protection inconvenient or data sharing convenient.

Second, the chapter highlights that privacy protection is made inconvenient through registering options (social login) in digital environments. The participants reported choosing these options because they found them more convenient than creating an account. Using these options may not be very conducive to privacy protection, but users are not given vital information about how their data will move, what permissions are included while signing in, and so on (Karegar et al., 2018). In other words, one is not pre-warned about the repercussions of using the Google or Facebook option (social logins) for registering. Moreover, as the participants in this study reported, these options were chosen because they were more convenient. Therefore, only the convenience of signing in through such options was made apparent. This enticement through convenience raises concerns. Waldman (2020) notes that dark patterns entice users to provide more information than they would if they were fully informed of the repercussions of doing so. According to Mathur et al. (2021), information hiding is an attribute of a dark pattern, and it includes concealing information from users. Since these social logins made it easier for the participants to share their data but hid essential information about the ramifications of choosing such options, they are dark patterns in digital environments that make data sharing convenient or privacy protection inconvenient.

10.5.1 Practical Implications

The commercial use of young people's online data is a central societal and policy concern (Livingstone et al., 2019b). Therefore, this study has practical implications for educators, regulators, designers, and corporations. Knowledge gaps related to various types of cookies, the function of cookies, user data protection mechanisms, and online data movements were discernible. Therefore, it is critical to increase young people's awareness of how their data move, especially their knowledge of cookies and available privacy protection mechanisms, as has also been recommended by other researchers (Stoilova et al., 2020; Holvoet et al., 2021).

Moreover, this chapter highlights that in contemporary digital spaces, privacy protection is made inconvenient first through lengthy and cryptic privacy terms, which deter young users from reading or editing them, thus leaving them with the more convenient option of accepting cookies. Researchers have emphasized that regulations should not just make it mandatory to seek users' consent but also impart clearer guidelines on how this consent should be gathered so that the process of consent-seeking becomes more meaningful and privacy-conducive (e.g., Utz et al., 2019) and dark patterns are avoided (Berens et al., 2024). Hence, these findings suggest that regulators must provide precise guidelines on how consent notices should be designed.

Lastly, this chapter highlights that privacy protection is not supported through registration options that entice users to resort to the more convenient option of using Google or Facebook without providing users with adequate information regarding the consequences of doing so. Researchers have called on designers to create ethical user interfaces, which include enhancing privacy protection and avoiding dark patterns (e.g., Gray et al., 2020). The findings suggest that designers and corporations must ensure that young users are provided with adequate information about the repercussions of using sign-in options instead of merely being shown the convenience of an option.

For companies that provide media services to young people, it should be part of their corporate social responsibility to attend to and respect young people's needs and interests, including their need for online privacy and safety (Lievens, 2018). However, by not adequately facilitating young people's privacy protection in digital spaces, corporations seem to be falling short of their social responsibility.

10.5.2 Avenues for Future Research

This study suffers from certain limitations, which must be highlighted. Young people's ability to protect their privacy may differ due to, among other factors, their technical knowledge and socio-economic backgrounds (Livingstone et al., 2019a). This study does not attend to differences *between* young people, which could have produced novel perspectives that are not covered in this study. Moreover, the participants were only from the Finnish capital region. Future studies should focus on young people from various regions in Finland and those belonging to different socio-economic backgrounds. Furthermore, the study only points out that young people felt privacy notifications are made incomprehensible on purpose. It does not include participants' design preferences for privacy notifications. Future studies could gather young people's views on how notifications should be designed.

AUTHOR'S NOTE

This work was supported by the Strategic Research Council (SRC) at the Academy of Finland (#352544).

NOTES

1. The language of the three FGDs in the bilingual school was English. The FGDs were conducted in Finnish in the other schools.
2. The guidelines for responsible conduct of research in Finland (TENK, 2019) were followed. Informed consent was obtained from participants 15 years and above. Parental consent was obtained for participants younger than 15 years. Pseudonyms have been used.

REFERENCES

Acquisti, A., Brandimarte, L., & Loewenstein, G. (2020). Secrets and likes: The drive for privacy and the difficulty of achieving it in the digital age. *Journal of Consumer Psychology*, *30*(4), 736–758. https://doi.org/10.1002/jcpy.1191.

Berens, B. M., Bohlender, M., Dietmann, H., Krisam, C., Kulyk, O., & Volkamer, M. (2024). Cookie disclaimers: Dark patterns and lack of transparency. *Computers & Security*, *136*, 103507. https://doi.org/10.1016/j.cose.2023.103507.

Braun, V., & Clarke, V. (2006). Using thematic analysis in psychology. *Qualitative Research in Psychology*, *3*(2), 77–101. https://doi.org/10.1191/1478088706qp063oa.

Eder, D., & Fingerson, L. (2002). Interviewing children and adolescents. In J. F. Gubrium & J. A. Holstein (Eds), *Handbook of interview research: Context and method* (pp. 181–201). Sage.

Finnish National Board on Research & Integrity (TENK). (2019). *The ethical principles of research with human participants and ethical review in the human sciences in Finland*. Finnish National Board on Research & Integrity (TENK).

Gibson, J. E. (2012). Interviews and focus groups with children: Methods that match children's developing competencies. *Journal of Family Theory & Review*, *4*(2), 148–159. https://doi.org/10.1111/j.1756–2589.2012.00119.x.

Gray, C. M., Chivukula, S. S., & Lee, A. (2020). What kind of work do "asshole designers" create? Describing properties of ethical concern on Reddit. *Proceedings of the 2020 ACM Designing Interactive Systems Conference*, 61–73. https://doi.org/10.1145/3357236.3395486.

Hollis, C., Livingstone, S., & Sonuga-Barke, E. (2020). Editorial: The role of digital technology in children and young people's mental health – a triple-edged sword? *Journal of Child Psychology and Psychiatry*, *61*(8), 837–841.https://doi.org/10.1111/jcpp.13302.

Holvoet, S., Jans, S. D., Wolf, R. D., Hudders, L., & Herrewijn, L. (2021). Exploring teenagers' folk theories and coping strategies regarding commercial data collection and personalized advertising. *Media and Communication*, *10*(1), 317–328. https://doi.org/10.17645/mac.v10i1.4704.

Horner, S. D. (2000). Using focus group methods with middle school children. *Research in Nursing & Health*, *23*(6), 510–517. https://doi.org/10.1002/1098-240X(200012)23:6<510::AID-NUR9>3.0.CO;2-L.

Karegar, F., Gerber, N., Volkamer, M., & Fischer-Hübner, S. (2018). Helping John to make informed decisions on using social login. *Proceedings of the 33rd Annual ACM Symposium on Applied Computing*, 1165–1174. https://doi.org/10.1145/3167132.3167259.

Keen, C. (2020). Apathy, convenience or irrelevance? Identifying conceptual barriers to safeguarding children's data privacy. *New Media & Society*, *24*(1), 50–69. https://doi.org/10.1177/1461444820960068.

Krämer, J., Schnurr, D., & Wohlfarth, M. (2019). Winners, losers, and Facebook: The role of social logins in the online advertising ecosystem. *Management Science*, *65*(4), 1678–1699. https://doi.org/10.1287/mnsc.2017.3012.

Lievens, E. (2018). Children's rights and media: Imperfect but inspirational. In E. Brems, E. Desmet, & W. Vandenhole (Eds), *Children's rights law in the global human rights landscape: Isolation, inspiration, integration?* (pp. 231–250). Routledge.

Livingstone, S. (2018). Children: A special case for privacy? *Intermedia*, *46*(2), 18–23.

Livingstone, S., Stoilova, M., & Nandagiri, R. (2019a). *Children's data and privacy online: Growing up in a digital age. An evidence review.* London School of Economics and Political Science.

Livingstone, S., Stoilova, M., & Nandagiri, R. (2019b). *Talking to children about data and privacy online: Research methodology.* London School of Economics and Political Science.

Mathur, A., Acar, G., Friedman, M. J., Lucherini, E., Mayer, J., Chetty, M., & Narayanan, A. (2019). Dark patterns at scale: Findings from a crawl of 11k shopping websites. *Proceedings of the ACM on Human–Computer Interaction, 3*(CSCW), 1–32. https://doi.org/10.1145/3359183.

Mathur, A., Kshirsagar, M., & Mayer, J. (2021). What makes a dark pattern... dark?: Design attributes, normative considerations, and measurement methods. *Proceedings of the 2021 CHI Conference on Human Factors in Computing Systems*, 1–18. https://doi.org/10.1145/3411764.3445610.

Milkaite, I., & Lievens, E. (2019). Children's rights to privacy and data protection around the world: Challenges in the digital realm. *European Journal of Law and Technology, 10*(1). https://ejlt.org/index.php/ejlt/article/view/674/912.

Srivastava, S., Wilska, T.-A., & Nyrhinen, J. (2023a). Children as social actors negotiating their privacy in the digital commercial context. *Childhood, 30*(3), 235–252. https://doi.org/10.1177/09075682231186486.

Srivastava, S., Wilska, T.-A., & Nyrhinen, J. (2023b). Awareness of digital commercial profiling among adolescents in Finland and their perspectives on online targeted advertisements. *Journal of Children and Media, 17*(4), 559–578. https://doi.org/10.1080/17482798.2023.2257813.

Stoilova, M., & Livingstone, S. (2022, September 8). CORE: A knowledge base on children & youth in the digital age. *Children and the Digital Environment.* https://core-evidence.eu/posts/digital-environment-takeaways.

Stoilova, M., Livingstone, S., & Nandagiri, R. (2019a). *Children's data and privacy online: Growing up in a digital age. Research findings.* London School of Economics and Political Science.

Stoilova, M., Livingstone, S., & Nandagiri, R. (2020). Digital by default: Children's capacity to understand and manage online data and privacy. *Media and Communication, 8*(4), 197–207. https://doi.org/10.17645/mac.v8i4.3407.

Stoilova, M., Nandagiri, R., & Livingstone, S. (2019b). Children's understanding of personal data and privacy online – a systematic evidence mapping. *Information, Communication & Society, 24*(4), 557–575. https://doi.org/10.1080/1369118X.2019.1657164.

Utz, C., Degeling, M., Fahl, S., Schaub, F., & Holz, T. (2019). (Un)informed consent: Studying GDPR consent notices in the field. *Proceedings of the 2019 ACM SIGSAC Conference on Computer and Communications Security*, 973–990. https://doi.org/10.1145/3319535.3354212.

Van der Hof, S. (2016). I agree ... Or do I? A rights-based analysis of the law on children's consent in the digital world. *Wisconsin International Law Journal, 34*(2), 409–445.

Van der Hof, S., Lievens, E., Milkaite, I., Verdoodt, V., Hannema, T., & Liefaard, T. (2020). The child's right to protection against economic exploitation in the digital world. *International Journal of Children's Rights, 28*(4), 833–859.

Waldman, A. E. (2018). *Privacy as trust: Information privacy for an information age.* Cambridge University Press.

Waldman, A. E. (2020). Cognitive biases, dark patterns, and the "privacy paradox". *Current Opinion in Psychology, 31*, 105–109. https://doi.org/10.1016/j.copsyc.2019.08.025.

11. Pay to play: the ideology of microtransactions and loot boxes in video games

Christopher McMahon

11.1 INTRODUCTION

The presence of microtransactions within video games alters the experience of play. The *FIFA* (*EA Sports FC*) series and *Fortnite* will be the focus of this chapter, with the *Call of Duty* franchise also being used to support the analysis of how microtransactions serve to reinforce neoliberal ideology and subjectivities by motivating players to alter their presence within the play-space. Frequently, this alteration is related to being able to perform better and to *appear* to be performing better. Illustrating how microtransactions function according to the form they take can reveal the underlying ideology behind them and how this reinforces negative entrepreneurial subjectivities. Such subjectivities relate to an understanding of neoliberal ideology that sees it as a way of organizing thought and action according to ideas of optimization and rationalizing input and output. From this, it can be considered what the possible consequences are for young people carrying out leisure in such an environment. Essentially, the position of this work is that microtransactions exist for the purpose of monetizing both positive and negative emotions that arise from digital gaming.

Microtransactions are in-game purchases available within a significant portion of video games. The items available for purchase via these interactions are virtual but (in some form) cost real money. The player could directly pay for an item with real currency or purchase in-game currency with money that is then used to purchase digital items. These digital items can or cannot affect gameplay. Some microtransactions could give the player items that give them an advantage in multiplayer games, or the transaction could be for a purely aesthetic item that alters the appearance of the player's digital avatar. For example, in *EA Sports FC 24*, players can purchase FC Points (at the time of writing, 1,050 FC Points are worth £8.99). These points can be used to buy packs of players that can ultimately improve the player's team. In *Fortnite*, the player similarly purchases V-Bucks (at the time of writing, 1,000 V-Bucks can be purchased from the Microsoft Store for £6.99). This virtual currency can be used to purchase skins or loot boxes. Loot boxes contain a range of potential in-game items, so there is a randomness to their purchase. As will be discussed, this is where trepidation and significant areas of concern arise from the use of such mechanics within video games enjoyed by millions of young people.

The key theoretical assumptions relevant to this work are drawn from Althusser and Foucault. The works of both these theorists are key for understanding neoliberal ideology as something that organizes thought and action around maximizing output, effectively aiming to compress time and distance (and in doing this experience) to arrive at an output in the fastest possible time. Althusser's notions of ideological and repressive state apparatus and interpellation (Althusser, 2014) can be used to understand the functionality of video games (and the mechanics within them) in reinforcing neoliberal ideology. As discussed in *The Corruption of Play* (McMahon, 2022, p. 102), the video game structure interpellates the player as a subject of the game's structures. The player's agency is acknowledged (and largely structured) by the video game structures themselves. The player will be directed towards various actions and considerations according to what the structure is. This can be developed by considering Foucault's work discussing the orientation of human action around neoliberal logics. Disciplining structures produce 'a positive economy; it poses the principle of a theoretically ever-growing use of time … it is a question of extracting, from time, ever more available moments and, from each moment, ever more useful forces' (Foucault, 1991, p. 154). Foucault notes that this is a body disciplined by authority (1991, p. 155). This occurs to the digital avatar of the player, and the player themselves in searching for additional rationalized meaning within leisure. Game structures govern and discipline action effectively as players are often beholden to their rules. In summary, the ideological context of the work is one in which agency is conducted according to market and economic imperatives and logics.

The title of this chapter is designed to function in various ways. The chapter's central argument concerning microtransactions is that they are orientated around eliminating and optimizing time and monetizing negative and positive emotions that result from a particular formulation of play. The proliferation of such mechanics within major video game series can result in understandings of play being grounded in economic means, becoming entrenched with the millions of young people that play them. In short, microtransactions offer a means to eliminate negative emotions that result from delayed or denied gratification or offer a means to enhance the potential positive outcomes of play. Microtransactions can be characterized as offering an enhanced form of play in this regard. The second function of the title is to note that the play itself is orientated around consumption rather than the game. Using in-game purchases is play within the contexts discussed in this chapter. The boundaries of the game can shift to a site of consumption that is enabled by the networked environment of modern video games. Be it the PlayStation store, the Microsoft Store, the Nintendo Store, Steam, or Epic Games, the player's space has a site of consumption operating within it.

The *FIFA* video game series began in 1993 and grew into a global phenomenon. Starting with 16-bit graphics, the series in its current form has photoreal depictions of football matches. It was reported on the BBC News website that the final entry in the *FIFA* series (*FIFA 23*) before it changed to *EA Sports FC* had a record-breaking launch, noting 10.3 million players in the first week. The series has changed to *EA Sports FC* following a reported disagreement between Electronic Arts and FIFA over the cost of the licence to use the FIFA name (Powell, 2023). The shift in branding does not appear to have slowed the series down, with Eurogamer reporting that *EA Sports FC 24* recorded around 11 million players in the first week after release (Wales, 2023). *FIFA* and *EA Sports FC* will be referred to interchangeably within this chapter unless stated otherwise. The Ultimate Team mode is the mode of the game in which microtransactions take place and the one which a substantial portion of the player base engages with.

Fortnite was released in 2017 and is nothing short of a media phenomenon. There are distinctive modes that have been introduced at various points during the development and evolution of the game. Despite rarely seeing precise player count information, it is predicted that *Fortnite* has an average monthly player count of 252 million, with a peak of 25 million in a single day. Most of the players are young people. Figures from 2018 indicated that 45 per cent of all young people (aged 8 to 17) had played *Fortnite* (Clement, 2023). Almost 1 in 2 young people in the US play the game. *Fortnite* is a free-to-play game. If you have hardware that the video game is available on, you can play it at no cost. All purchases are optional, and they are deemed as having only a cosmetic effect on the game, such as changing the appearance of the player avatar.

11.2 THEORETICAL APPROACH: MICROTRANSACTION AS FORMAL STRUCTURE

Before continuing, it is important to apply nuance to this discussion in whatever ways possible. Especially when considering the impacts upon young people, there is the potential to fall into the position of a moral panic reminiscent of violent video games in the 1990s (Markey & Ferguson, 2017). It would be an overstatement to frame all forms of in-game purchases as inherently predatory, but there is an inclination towards this when children and young people are a core part of the audience for video games that actively push their in-game purchases. The compulsive nature of microtransaction purchases has been acknowledged previously, following the emergence of a moral panic as gaming became framed as an addiction due to this compulsive nature (McMahon, 2022, p. 124). It was reported by BBC News that a gaming disorder has been recognized (Rajan, 2018) and behavioural experts cited in an article in *USA Today* have likened *Fortnite* to heroin (Haller, 2018). Therefore it is justified to be sceptical of such a framing of play in video games and to argue that the moral panic is the result of an inability, or lack of will, to identify the extent to which play and the digital spaces in which it can occur are held hostage by capital and neoliberal subjectivities. This is the approach and position from which microtransactions are critiqued here. This work will not be able to sufficiently comment on the regulations and legality of the mechanics or to speculate on the true intentions of developers and publishers within the scope of this work. There is excellent and extensive research on the legality, regulation, and policy relating to microtransactions (such as Xiao's (2023) work and his documentation of such approaches). Additionally, this chapter will not engage with particular individuals' use of the mechanics. When specific cases of microtransactions are used within this work, they operate as a case study and point of consideration. This work engages in a way that considers what forms the microtransactions take and their potential ideological implications rather than suggesting their ban or restriction within video games or issuing a categorization of those that use them, frequently or infrequently.

Research on individuals' microtransaction use has revealed some troubling trends. The conclusion of Close et al. (2021) notes that developers appear to be profiting from mid to high-risk gamblers, with spending patterns mirroring those within gambling. Internet Gaming Disorder (IGD) is linked to free-to-play games in a study that also notes that the design of such games 'transports the feeling of self-efficacy', with spending mechanics allowing players to navigate issues related to it (Dreier et al., 2017, p. 332). However, King et al. (2020) observe that spending practices are likely influenced by peer practices and not ideas of a gaming disorder. These sorts of observations are concerning, but it can be argued that the primary concern within this issue is potentially exploitative structures rather than the practices within them.

Characterizing play and forms of gaming as addictive is problematic. It is difficult to simply say that the use of microtransactions constitutes a gambling problem or a disorder. It personalizes what is likely to be a structural issue. The answer is likely a more complex, deeply rooted ideological issue that incorporates decades of economic, social, and cultural forces. What this work considers is that in a socio-economic system that often seeks to maximize profit at the expense of other considerations, it stands to reason that feelings and forms of action will be harnessed towards the end of capital accumulation. Leisure is held captive by capital, so people should not be characterized as ill in a society that has driven them to find means of enjoyment and happiness and are caught within the webs they are trying to escape.

After the Columbine shooting in 1999, there was a significant increase in research on violent video games (Markey & Ferguson, 2017, p. 103). Markey and Ferguson note how the event set off a quest for answers and that there were questions asked concerning the influence of goth music culture, the film *The Matrix* (Wachowskis, 1999), and video games such as *Doom* (1993) (Markey & Ferguson, 2017, pp. 103–104). A similar phenomenon is observable in relation to microtransactions as we try to explain the social ills that exist around them. Microtransactions exist within an extensive media network and ecology of everyday life. For those playing the football simulation in *EA Sports FC 24*, they will see the gambling sponsors on the shirts of multiple top football clubs and, in physical reality, see advertising boards and sponsorships for major gambling companies. Mediated forms show symptoms of a society that is orientated around maximizing capital.

As stated at the start of this section, this topic should be approached with nuance. The inclusion of microtransactions within the video game medium is different from the inclusion of violence, and by no means are microtransactions being given a pass by relating their discussion to a panic. Microtransactions can be identified as *bad*, in a way that media violence cannot and should not be so identified. The aim of highlighting and discussing the material so far is to reflect upon our contextual understanding and why the argument for a neo-formalist approach to microtransaction mechanics can be theoretically useful in envisioning why and how they are used. For example, the noted influence of microtransactions and spending on self-efficacy in Dreier et al.'s (2017) work is what this approach aims to build on. From such an observation, it can be seen that microtransactions do arguably seek to exploit feelings that are a result of play with a win–loss condition. Winning and losing is a natural element of (most) play and games, with video games such as *Fortnite* and *FIFA* being prevalent examples of having winners and losers. The neoliberal organization of technology is seen in how the feelings that result from winning and losing are monetized; individuals may engage with microtransactions to heighten or limit these feelings. The technology and media form itself is what is analysed within this chapter rather than any individual's use of them.

The approach to microtransactions in this chapter is informed by neoformalism. The microtransaction itself is considered a formal component of the video game and its structure, just as filmic or textual elements would be in cinema. In relation to cinematic poetics, Bordwell states that 'meaning, structure, and process – these three aspects of any representational system are also central to poetics' (1989, p. 376). The player, like the viewer of a film, has a certain understanding of how their engagement with the medium will unfold. In this regard, the meanings of the text, its formal structure, and the very process of play are highly significant. Microtransactions can be made to feel like a natural element of play rather than a formal attachment to it, so this work will analyse it as a natural formal element in order to highlight its effects. Thompson notes how a neoformalist approach looks for the *dominant*, which is 'the main formal principle a work or group of works uses to organize devices into a whole' (1988, p. 43). In this instance, the microtransaction is recognized as the dominant factor of the medium. From this, its effects can be analysed and considered for ideological significance. The notion of defamiliarization that emerges from such an approach is key. Thompson notes this as similar to works of art accomplishing a renewing effect on our 'mental processes', and the term connotes the 'basic purpose of art in our lives' (1988, pp. 10–11). This process can also function in a way that reproduces ideological positions and subjectivities, such as a neoliberal notion of the self and assessment of thought and action. Microtransactions achieve an effect of defamiliarizing play and playful activity, orientating a play-space and player agency around notions of optimization. The intended point of analysis here will not benefit from testimony from spectators or players within the present moment, rather the general audience assumptions within the context of formal and aesthetic structures are key. By discussing elements of the form and reaction, areas of significance can be highlighted. The purpose of this is to consider the presence of such mechanics within a play-space, considering how acts of consumption react to and influence acts of play, and how they can become one and the same.

11.3 CONSUMPTION IS IN THE GAME: MICROTRANSACTIONS IN *FIFA/EA SPORTS FC*

This section discusses how microtransactions in the *FIFA* and *EA Sports FC* series are able to monetize the positive and negative emotions that result from engagement with the video game. The Ultimate Team mode in the *FIFA* series reflects a desire for instant gratification, and allows for the meaningful experience the game offers to be bought rather than arrived at via play (McMahon, 2022, pp. 125–128). Additionally, microtransactions offer a means to circumvent negative emotions by providing a means to improve your chances of winning simulated football games and by providing an alternative means of play by continually opening *FIFA* packs.

The obvious needs to be stated sometimes: the *FIFA* series are games. A game and its rules can be stressed, exploited, and broken with a mastery of relevant techniques and an understanding of the rules that govern play. This is especially true for video games where a thorough knowledge of mechanical functions and AI operations can allow a player to accurately predict or manipulate how the game itself behaves. This produces especially strong feelings of negativity in relation to the *FIFA* games, if a player is seen to be exploiting or noticeably stressing game rules, mechanics, and physics. It can be assumed that the player inflicting a negative experience upon another player in online multiplayer versions of the game is themselves having a positive experience. This is further reinforced by how the *FIFA* games allow players to celebrate scoring with elaborate celebrations that require a specific input of the controls to execute, implying a deliberateness in the choice of celebration. In this regard, the *FIFA* video games can be ruinous or lucrative in terms of emotional reward or loss, but that is the risk and that is the reason for simulating football. It can be argued that this needs to be accepted as a natural component of play (you win some, you lose some), but the networked nature of digital video games complicates this. The play-space is porous, as the infrastructure that hosts and enables the play is also a site of consumption and content sharing. Neoliberal subjectivities are fostered in the networked space of AAA (video games produced and/or distributed by major publishers, such as EA) video games, so players will seek to heighten and intensify gains and remove loss and unpredictability (McMahon, 2022). This will be the drive to limit or negate negative emotion and to heighten and multiply positive emotion.

FIFA games have a strong interconnectivity with other media formats and platforms. EA themselves have the EA Creator Network (Electronic Arts, n.d.), an initiative that demonstrates the significant role content creators play in the circulation of networked video games. Beyond officially sanctioned content shared on platforms such as YouTube and Twitch, *FIFA* rage videos and long rants about the state of the game from professional or notable players have become a genre of video in their own right. A video of a popular streamer and YouTuber reacting to a pro-player complaining about the state of the game in *EA Sports FC 24* is indicative of this form of content (NickRTFM, 2023). Within the video, a player is lamenting the ease with which players can force a goal by deploying a particular strategy of running to the byline and cutting the ball back across the penalty area. The player also laments that the way to defend against this is to play extremely negative football, known as *parking the bus*. If both players do this, games could be very negative, frustrating, and formulaic. EA has even banned the tactical formations related to this issue (three and five back formations, forcing teams to play with four defenders) at pro-player events to make the games more exciting (Perry, 2023), signalling a competitive flaw within their own game. Negative games produce negative emotions, and a way to address this is to encourage the better players to execute a play more consistently within the game mechanics or find play within another area of the game.

As indicated in previous research (McMahon, 2022, pp. 125–127), opening packs in the Ultimate Team mode is a form of play in itself. These packs can be acquired with in-game currency (which can be earned via play or purchased with real money). As with rage and rant videos, pack opening videos are an extensive genre in their own right, with notable YouTubers and streamers devoting multiple videos to just opening packs. Additionally, the more packs the player opens, the more a chance they have of receiving a virtual footballer that can help them win more games, or they can trade virtual footballers for in-game currency in order to buy another virtual footballer (or more packs) from an internal trading market. A sense of elation upon unpacking a virtual footballer that the player desired can be just as great if not greater than achieving victory against other players online. This feeling can be continually pursued with the use of microtransactions. With this framing, comparisons between gambling and loot-box-style mechanics in video games arise. The player is effectively sitting at a digital slot machine, hoping their next pull will be better than the last.

To summarize, *FIFA* players experience both positivity and negativity within the simulated football match and in the purchasing of packs in the Ultimate Team game mode. Microtransactions enable the heightening and quickening of gratification and can help alleviate the potential for negative emotions. Even if the literal playing of a simulated football game is not engaging the player, consumption has the potential to more than make up for the lack of a meaningful experience. Play is defamiliarized to the point where financial transactions resemble playful activity. Agency is exercised most effectively when consuming, and the game offers players rewards when they consume, potentially penalizing them when they do not.

11.4 DUBS, DROPS, AND DRIP: THE NECESSITY OF NON-ESSENTIAL MICROTRANSACTIONS IN *FORTNITE*

This section discusses how *Fortnite* necessitates an engagement with premium cosmetics in the game to offer a substantively meaningful experience to the player. It is difficult to overstate the success of *Fortnite* and the apparent vice-like grip it has on young audiences. Whilst it is difficult to offer a catch-all reason for the appeal and success of media (especially video games) due to the myriad ways experiences can be had, there are strong identifiable themes that can be closely analysed. Carter et al.'s (2020) work accurately identifies a key appeal of *Fortnite* in the 'intersections with YouTube and game livestreaming, the way the game acts as a vehicle for social capital and the performance of identity, and the rich sociality of play' (Carter et al., 2020, p. 453). In considering the force motivating the purchasing of microtransactions in *Fortnite*, we can consider the following argument:

> It is not just the content creators that create value and qualify different goods (skins and emotes) and services (battle passes), but the nonhuman algorithmic

> recommendation systems and third-party analytics that shape, encourage, and constrain the content that they produce for the platform. Fortnite creates content for YouTubers to monetize, and YouTubers create value for Fortnite to monetize. The game-driven audience drives the popularity of Fortnite on YouTube, encouraging more streamers to produce Fortnite content, cascading and sustaining its popularity over other games. (Carter et al., 2020, p. 465)

Fortnite, like *FIFA* and *EA Sports FC*, is part of a broader network. Young people engage with media within a converged environment, as the media titles serve as vehicles for content across various formats. *Fortnite* is no different, as the game will have a significant impact on content-sharing sites such as YouTube and social media sites that serve as communication channels. To play *Fortnite* is to actively take part in social life, either via communication with others or by creating your own content. *Fortnite* creates an issue when microtransactions are introduced. It is effectively the monetization of social and cultural everyday life.

Neely's work engages with the ethics of microtransactions themselves, acknowledging that they can take different forms and serve different functions. Neely ultimately argues that 'fixed cosmetic rewards are ethically permissible, random rewards of all types are ethically problematic, and fixed functional rewards can be acceptable, but only under certain conditions' (2021, p. 228). Neely outlines cosmetic items as things that do not affect gameplay, and elaborates upon ethical issues that arise from their use by stating:

> Ethical issues can still arise from the sale of fixed cosmetic rewards. For instance, designers need to consider how they are marketed, and it could be possible to market them to child players, say, in a predatory way. There could also be ethical concerns about the pricing of items, though designers that try to charge too much may simply find that players are unwilling to purchase them. But these sorts of ethical issues are related to sales and marketing techniques in general rather than being specifically tied to microtransactions. (Neely, 2021, p. 236)

This assessment is accurate but only when play is defined as the actual conduct of the game within strict terms and parameters. Play that operates within networked contexts needs a broader definition, one that encompasses the functionality of play in the everyday life of young people. As highlighted by Carter et al. (2020), *Fortnite*'s appeal is found within its intersections with other media as well as its function in generating social capital and in identity formation. This can be extended to being the case for most, if not all, networked video games. The previously discussed *FIFA* series will be doing the same, serving a significant function in how young people think about who they are and what they engage with beyond the play of the game itself. Consuming cosmetics is play and socialization.

It has been highlighted that there is a stigma to playing *Fortnite* while having the avatar equipped with a default (and hence free) skin, indicating the player has spent no money on V-Bucks or skins. Hernandez reported in Polygon, a source of gaming news, regarding bullying that related to *Fortnite*, that:

> One student in Towler's class 'begged his parents for [money] to buy a skin because no one would play with him' because he wore basic virtual clothes. While the bullying wasn't always Fortnite-specific, Towler recalls that it seemed 'vicious for [the student] to have another avenue for the meaner kids to attack him.' Things got better for that kid, but when your social scene begins and ends with Fortnite, having nobody to play with is like a mark of death. (Hernandez, 2019)

When the functionality of the playing of a video game is connected to social life, it will feed into the various dynamics within it. Whether you are concerned about bullying, wish to shield yourself from it, or if you are the bully, the purchasing of microtransactions is the social pretext through which this can be exercised. As Hernandez noted, a young person was begging for money to alter how their play of the game was occurring. The instinct was not to stop playing (nor should it be); it was to remedy social ills via the avenues that are left open to do so – spending money. A link was established between peer purchasing practices and in-game purchases in *Fortnite*, with a study noting that spending was

> related mainly to social influences … spending behaviors may have important social underpinnings, and that these social factors appear to be more pertinent in predicting spending for some players than variables that are commonly related to individual differences in problem gaming studies. (King et al., 2020, pp. 5–6)

Rather than a specific behavioural issue of players, the structures foster a particular form of subjectivity and action. Agency is orientated around establishing yourself in an optimized way in relation to your peers.

The argument and understanding of cosmetics not affecting gameplay becomes more complicated when we think critically about what happens during the play of games such as *Fortnite* and other battle royale games, such as *Call of Duty: Warzone* and *Call of Duty: Warzone 2.0*. It has been a persistent notion within gaming culture that different skins within *Fortnite* offer a distinct advantage, although this has been denied by Epic Games and it is true that different skins have the same hitbox[1] (Sabarwal, 2023). However, elite-level players continually favour slimmer cosmetic skins, despite them having the same hitbox as significantly larger skins. An article explains that this is due to the avatar with such a skin taking up less screen space (Sabarwal, 2023). Upon reflection this appears to be a logical choice if you seek a competitive advantage. While the hitbox may well be the same, when a player's avatar is on the screen of opposing players, they will appear to be a smaller target and will better fit behind objects that may delay them entering the sight of opposition players. The cognition of opposing players seeing the avatar as a smaller target will affect their conduct and hence offer an advantage (a small advantage is still an advantage) to the player who appears harder to hit and finds it easier to hide.

The issues discussed in this chapter are not unique to *FIFA* and *Fortnite*. A skin available in the online multiplayer mode of *Call of Duty: Modern Warfare 2* and *Call of Duty: Modern Warfare 3* has caused 'havoc' for players (Kennedy, 2023). The Gaia cosmetic made some players hard to see, offering a distinct advantage to players with the skin equipped. The advantage was so pronounced that Sledgehammer Games ultimately removed the cosmetic item from the game temporarily while they adjusted it. One player was quoted as saying:

> In every map I played, I could barely see players who were using this stupid skin, how is that fair and why no almost no one talks about it? It should be a removed from the game as it promotes [pay-to-win] characteristics and its literally invisible at first glance. (Quoted in Kennedy, 2023)

Here, the player identifies a connection between the nature of microtransactions and the distinct advantages some cosmetic items can offer. Even if not intentional by the developers, cosmetic differences in video games that depend partially on the perception of other players' avatars will alter the conditions of the game.

Additionally, it is difficult to frame cosmetic items as purely cosmetic even when you are not thinking about the actual playing of the central game. Separating cosmetic items like this from play relegates the significance of aesthetics themselves. If a primary appeal of *Fortnite* is the integrity it has to broader social functions and identity formation, cosmetic items are an essential component of the gameplay itself. Similar to how pack openings in *FIFA* are a spectacle, play is occurring via consumption. In such instances as when a particular skin can provide a competitive advantage or when a particular cosmetic set of items signals a lack of various forms of capital, there are significant impacts upon gameplay when we broaden the idea of what play is and where this play is occurring.

To summarize, cosmetic skins accessed via microtransactions offer the player a heightened experience within *Fortnite*. Not only do they potentially offer an advantage within the competitive playing of the game itself, but they also assist in generating social capital. Play is orientated around ideas of accumulation using the formal structures of the game and how microtransactions assist the player in reaching the assumed good outcomes of their play.

11.5 DISCUSSION: PAYING TO PLAY AND PAYING AS PLAYING

Ultimately, the negative emotions that result from the formation and structure of major digital video games have been monetized, along with the elements of joy that reside at the core of the experience of play. Whether the player is obtaining their intended experience, or they wish to alter the conditions of the game to move them closer to their intended experience, microtransactions offer the ability to arrive at the intended outcome sooner. Microtransactions have not caused the state of digital video games; rather, they are an exploitation of the feelings that were already there. The networked nature of video games constructs an environment where the young people playing them are all too aware of their economic constraints and how these constraints limit their agency within the play-space. The presence of in-game mechanics does exacerbate these issues significantly, but they are not the root of the problem. For all intents and purposes, consumption has become a form of play within the context of networked digital games. To summarize the general logic of using microtransactions within the two video game series this chapter has analysed, *FIFA*'s microtransactions capitalize on frustration and elation and *Fortnite*'s tap into needs for social capital and a sense of belonging. This chapter's analysis does not suggest a call for a banning or restricting microtransactions within video games. A case for that may very well be made; rather, this chapter has sought to highlight the underlying ideology that drives players to make use of in-game purchases. Therefore, it can be argued that it is a culture of immediacy, exacerbated by the reproduction of neoliberal subjectivities within AAA video games, that fosters microtransaction use. Entrenched ideological positions are not something we can ban or restrict players from engaging with in a *neoliberal* way. The hobbies of the young are held hostage by capital, and their social energies are monetized. A major channel through which these energies are directed is video games. In this sense, microtransactions are exploitative of young people in the way that broader structures are exploitative. It is difficult to think of how to proceed in addressing this beyond a fundamental restructuring of digital infrastructures and diversifying ownership models and production cultures within the entertainment industries. Asking young people to stop playing the sorts of games discussed in this chapter is not the answer. A hopeful note to consider would be that play is given a sacred place within the lives of young people, and that the emotional highs and lows of gaming, which microtransactions seek to exploit and monetize, become an accepted part of play rather than something that can be limited or enhanced with in-game purchases.

11.5.1 Practical Implications

The main implication of such theoretical investigations of microtransactions is to raise awareness of the how and why of microtransaction consumption. It is a chance for reflection on the framing of potentially problematic microtransaction use, as well as allowing an opportunity for those who use microtransactions in their play to consider why. As noted, a way to address the issues highlighted in this chapter is with broad-reaching cultural, social, ideological, and regulatory changes. This is easier said than done and will need to be informed by a broad range of research and debates. For the individual player, the process of defamiliarization offers the opportunity to interrogate the seemingly automatic and common-sense process of engaging or not engaging with microtransactions in their different forms. Considering why you would purchase something within certain types of games or whether they involve cosmetics or time-saving devices could reveal underlying motivations. Happily engaging with the purchase of cosmetic items but finding the purchase of time-saving devices (such as XP or currency boosts) uneasy may suggest a more nuanced understanding of when microtransaction use is motivated by something other than player enjoyment.

11.5.2 Avenues for Future Research

Building upon the previous section, an opportunity for future research would be an investigation of player microtransaction use that utilizes the process of defamiliarization. Defining microtransactions in two broad categories (cosmetic and time-saving) and interviewing players on the use of which one they would not normally use could reveal insights into the underlying assumptions that motivate use. Additionally, further theorization can be conducted in relation to the presence of microtransactions within mobile gaming alongside PC and console gaming. Due to the various cultural and social assumptions of who games on what platform, microtransaction use will likely be understood and embraced in varying ways by active users of each platform.

NOTE

1. A hitbox is the area on an avatar where an attack towards it will be effective and registered as a hit in some form.

REFERENCES

Althusser, L. (2014). *On the Reproduction of Capitalism: Ideology and Ideological State Apparatuses*. London: Verso.

Bordwell, D. (1989). Historical poetics of cinema. In R. B. Palmer (Ed.), *The Cinematic Text: Methods and Approaches* (pp. 369–398). London: AMS Press.

Carter, M., Moore, K., Mavoa, J., Horst, H., & Gaspard, I. (2020). Situating the appeal of Fortnite within children's changing play cultures. *Games and Culture*, 15(4), 453–471.

Clement, J. (2023, August 25). Share of children who play Fortnite in the U.S. 2018. Retrieved from Statista: https://www.statista.com/statistics/985792/fortnite-playing-children-united-states/.

Close, J., Spicer, S. G., Nicklin, L. L., Uther, M., Llyod, J., & Llyod, H. (2021). Secondary analysis of loot box data: Are high-spending 'whales' wealthy gamers or problem gamblers? *Addictive Behaviors*, 117, 106851.

Dean, G. (2023, November 21). The 50 best Fortnite skins. Retrieved from PCGamesN: https://www.pcgamesn.com/fortnite/fortnite-skins-outfits-gliders-emotes.

Dreier, M., Wölfling, K., Duven, E., Giralt, S., Beutel, M., & Müller, K. (2017). Free-to-play: About addicted Whales, at risk Dolphins and healthy Minnows. Monetarization design and Internet Gaming Disorder. *Addictive Behaviors*, 64, 328–333.

Electronic Arts. (n.d.). FAQs. Retrieved from EA Creator Network: https://creatornetwork.ea.com/faq.

Foucault, M. (1991). *Discipline and Punish: The Birth of Prison*. London: Penguin Books.

Haller, S. (2018). 'This game is like heroin': Fortnite addiction sending kids to gaming rehab. Retrieved from *USA Today*: https://eu.usatoday.com/story/life/allthemoms/2018/12/09/fortnite-addiction-sending-kids-gaming-rehab/2221149002/.

Hernandez, P. (2019, May 7). Fortnite is free, but kids are getting bullied into spending money. Retrieved from Polygon: https://www.polygon.com/2019/5/7/18534431/fortnite-rare-default-skins-bullying-harassment.

Kennedy, V. (2023, November 21). Modern Warfare 2's 'Groot' skin continuing to wreak havoc in Modern Warfare 3. Retrieved from Eurogamer: https://www.eurogamer.net/modern-warfare-2s-groot-skin-continuing-to-wreak-havoc-in-modern-warfare-3.

King, D., Russell, A., Delfabbro, P., & Polisena, D. (2020). Fortnite microtransaction spending was associated with peers' purchasing behaviors but not gaming disorder symptoms. *Addictive Behaviors*, 104, 1–7.

Markey, P., & Ferguson, C. (2017). Teaching us to fear: The violent video game moral panic and the politics of game research. *American Journal of Play*, 10(1), 99–115.

McMahon, C. (2022). *The Corruption of Play: Mapping the Ideological Play-Space of AAA Videogames*. Bingley: Emerald Publishing.

Neely, E. L. (2021). Come for the game, stay for the cash grab: The ethics of loot boxes, microtransactions, and freemium games. *Games and Culture*, 16(2), 228–247.

NickRTFM. (2023, October). Reacting to 'EA NEEDS TO FIX THIS GAMEPLAY OR WE'RE FINISHED!'. Retrieved from YouTube: https://www.youtube.com/watch?v=kkfcQg5K-ic&ab_channel=NickRTFM.

Perry, K. (2023, September 28). EA FC 24 pros overjoyed as devs ban 3 & 5-back formations. Retrieved from Dexerto: https://www.dexerto.com/ea-sports-fc/ea-fc-24-pros-overjoyed-as-devs-ban-3-5-back-formations-2313490/.

Powell, S. (2023, September 27). Fifa to EA Sports FC: Name change is big gamble for UK's best-selling game. Retrieved from BBC: https://www.bbc.co.uk/news/entertainment-arts-66917614.

Rajan, A. (2018, June 18). Gaming becomes the latest addiction. Retrieved from bbc.co.uk: https://www.bbc.co.uk/news/entertainment-arts-44518982.

Sabarwal, R. (2023, July 25). Fortnite hitboxes: Do skins have a competitive advantage? Retrieved from Dexerto: https://www.dexerto.com/fortnite/fortnite-hitboxes-do-skins-have-a-competitive-advantage-2222816/.
Thompson, K. (1988). *Breaking the Glass Armor.* Princeton, NJ: Princeton University Press.
Wachowskis (Directors). (1999). *The Matrix* [Motion Picture].
Wales, M. (2023, October 9). EA Sports FC 24 trumpets 11m players in first week, 1m more than FIFA 23. Retrieved from Eurogamer: https://www.eurogamer.net/ea-sports-fc-24-trumpets-11m-players-in-first-week-1m-more-than-fifa-23.
Xiao, L. (2023). Opening the compliance and enforcement loot box: A retrospective on some practice and policy impacts achieved through academic research. *Societal Impacts*, 1(1–2), 100018.

Ludography

Call of Duty: Modern Warfare 2 (2022). Developed by Infinity Ward. Publisher Activision.
Call of Duty: Modern Warfare 3 (2023). Developed by Sledgehammer Games. Publisher Activision.
Call of Duty: Warzone (2020). Developed by Infinity Ward, Raven Software. Publisher Activision.
Call of Duty: Warzone 2.0 (2022). Developed by Infinity Ward, Raven Software. Publisher Activision.
Doom (1993). Developed by id Software. Publisher id Software.
EA Sports FC 24 (2023). Developed by EA Sports Vancouver, EA Sports Romania. Publisher EA Sports.
FIFA 09 (2008). Developed by EA Canada. Publisher EA Sports.
FIFA 16 (2015). Developed by EA Canada. Publisher EA Sports.
FIFA 19 (2018). Developed by EA Sports Vancouver, EA Sports Romania. Publisher EA Sports.
FIFA 23 (2022). Developed by EA Sports Vancouver, EA Sports Romania. Publisher EA Sports.
Fortnite (2017). Developed by Epic Games. Publisher Epic Games; Warner Bros. Interactive.

12. The gig economy in the digital era: platform work as a crutch for precarity?

Julia Nuckols

12.1 INTRODUCTION

In recent decades, a shift in the political economy of labour markets has transpired, leading to an increase in less traditional forms of employment (Watson et al., 2021; Collier et al., 2017). Informal employment has become particularly commonplace via digitalization (Muntaner, 2018), with young people typically engaging in these means of work (Huws et al., 2017). Statistics from the United States of America have also found non-whites to be more likely to partake in gig work, while in European countries, it is migrants (Anderson et al., 2021; Sutela, 2023). However, these forms of informal employment are also believed to cause precarity in one's occupational endeavours, as gig work is often conducted by groups that are already considered societally vulnerable, such as young people or immigrants (Vyas, 2021; MacDonald & Giazitzoglu, 2019). The digital shift in the area of gig work primarily involves app-based services that consist of both consumer users as well as worker users – that is, those who require a service and those who facilitate said service in a freelancing nature (Florisson & Mandl, 2018). Of course, other means of freelancing in the digital realm exist, such as content creation, where typically a social media user can facilitate promotions for brands (Arriagada & Bishop, 2021). However, this chapter will focus solely on digital gig work facilitated by platform services such as Uber or DoorDash, which will be referred to as platform work.

While traditional forms of employment typically entail set working hours, benefits, and agreements of continuity, informal employment does not usually come with these attributes. Instead, it typically involves varying contract durations, unpredictable working hours, and a lack of benefits (Woodcock, 2020; Muntaner, 2018). The low threshold for entry into informal work may make it desirable, or in some cases a necessity for making ends meet. It also offers solutions for those in search of solely part-time work or who require or desire flexible working hours (Ly, 2020).

Research has found that on a global scale, younger age is a typical attribute of those partaking in platform work, as well as being a migrant or non-white (Anderson et al., 2021; International Labour Organization [ILO], 2021; Sutela, 2023). Given how platform jobs can provide flexibility and low-threshold entry, they may be desirable for younger individuals as these job qualities are typical for part-time work and ideal, for example, for students (Ray & Pana-Cryan, 2021). Younger age is also commonly associated with uncertainty, including from a financial and career perspective, therefore adding more incentive for younger individuals to seek such work, especially if they are in a situation where other means of work and income are unavailable (Graham et al., 2017; Ranta, 2015). However, this raises the question of whether such work opportunities also come with risks for young people.

This chapter aims to offer an introduction to informal employment driven by platform work as part of the modern gig economy. The next section of this chapter will begin by elaborating on the terms *platform work* and *gig economy* while also providing practical examples of various forms of platform work. This is followed by a historical overview of how platform work has coincided with general shifts in work culture in modern times. This will be followed by an examination of the pros and cons of platform work, particularly in terms of young people, with the pandemic discussed as a practical example of the risks and opportunities present in platform work. The chapter concludes by summarizing what has been covered and what are the implications for policymaking and academic research.

12.2 DEFINING GIG WORK IN THE PLATFORM ECONOMY CONTEXT

Informal work generally refers to work that is outside of traditional employment which includes established continuity, set working hours, and employee benefits (Pfau-Effinger et al., 2010). *Gig work* comes under this banner, in which informal employment and labour takes place (Woodcock, 2020). *Gig workers* are individuals who are working under all different types of work in the gig economy, whether digital or not (Woodcock, 2020). *Platform workers* is the term commonly used to describe those who work via a mediating platform, such as courier service apps or transportation apps (Florisson & Mandl, 2018). However, this chapter will solely focus on platform workers via digital platforms, and therefore when using the terms *platform work/workers*, or *gig work/economy* it is in reference to work being done via established platform work services.

There are numerous different forms of platform work. There is delivery work, which involves the transportation of goods between businesses and clients by car, bicycle, motorbike, or scooters (Lord et al., 2023; Woodcock, 2020). Examples of apps or platforms that facilitate work of this nature include Wolt, Uber Eats, DoorDash, and Instacart (Griesbach et al., 2019). Their model is based on an app or platform serving as a mediating entity, connecting those who deliver to those who order from businesses.

Another common form of platform work involves taxi and driver services, which are facilitated via online platforms rather than more traditional means (e.g., being stationed at a stop while waiting for customers) (Davidson et al., 2023; Woodcock & Graham, 2019). Even though there are more formal means of taxi work in which drivers work directly under the supervision and responsibility of a company, more drivers seem to be either working as independent freelancers or via applications or platforms such as Uber (Woodcock, 2020). The model of Uber and similar driver service apps is such that an individual with a vehicle may enlist themselves as drivers on the app and then select transportation orders by customers (Prassl & Risak, 2015). Other types of gig work via digital platforms include accommodation leasing (Airbnb) and various chores and tasks (TaskRabbit) (Bosma, 2022; Prassl & Risak, 2015). Airbnb, for example, functions so that an individual may list their properties for short-term accommodation for consumers, while TaskRabbit offers a place for consumers to list requests of various tasks and then have workers accept or decline said requests (Bosma, 2022; Prassl & Risak, 2015). Some research suggests that a division can be made between platform work that involves manual labour (rideshare, chores) for income purposes, and platform work that involves the selling or leasing of commodities (Airbnb, online markets) to enrich existing wealth (Hoang et al., 2020).

12.3 GIG WORKERS AND THE GIG ECONOMY – THEN AND NOW

A good example of traditional means of gig work would be freelance musicians who partake in occasional performances without knowledge of future bookings of their talent (Woodcock & Graham, 2019). The success and repetition of opportunities to perform are unknown, as is in what form they will receive payment if any payment is to be received (Woodcock & Graham, 2019).

The above is one of many situations in which there is typical gig-economy-related precarity involved in one's means of making a living. Gig work is, nearly without exception, characterized as work that does not include prolonged or indefinite contracts and is unpredictable (Woodcock, 2020). It is also, however, often accompanied by flexibility, low-threshold opportunities to work, and the ability to terminate contracts without the legal requirements associated with more formal working contracts (Roy & Shrivastava, 2020; Harmon & Silberman, 2019). It is with these characteristics that platform work largely falls under the definition of gig work (Woodcock, 2020; Vallas & Schor, 2020).

Work is an ever-changing construct that is an increasingly popular topic of research and debate, particularly in the wake of platform work. These changes have been rapid, with the platform economy suddenly becoming a solid part of work discourse (Watson et al., 2021; Muntaner, 2018). The types of formal employment that are now considered more typical are also relatively new and emerged primarily in the post-Second World War era. Some scholars refer to this as the "standard employment relationship," which emerged after the Second World War in an attempt to provide workers with jobs that were stable, socially protected, dependent, and full-time, and even then only in the West (Woodcock & Graham, 2019). These foundations in formal employment aimed to mitigate employee risks and provide security to the labour force. It is with this term "standard employment relationship" that employment of this quality is deemed normal today, making informal employment, such as platform work, understood as something unusual.

Despite informal employment being considered unusual, it is believed that precarious work of this nature has existed as long as the concept of trading one's time in return for a salary has (Kalleberg, 2009). In fact, much of the work prior to the establishment of formal employment was typically uncertain or unstable. Some believe that the emergence of the gig economy in its new popularity may be attributed to the increasingly competitive and unstable labour market of modern days (Mehta, 2023; Woodcock, 2020). In this light, with the low-threshold entry into work and the opportunity for companies to cut operating costs by only having employees complete jobs on demand, the gig economy in its modern context seems reasonable. The gig economy and, by extension, companies that provide platform work, also provide employers with less responsibility over the well-being of their labourers, making precarious work a lucrative option in more than just the fiscal sense (Harmon & Silberman, 2019). As stated by the International Labour Organization (ILO, 2011), one of the key definitions of precarious work is "in the most general sense, precarious work is a means for employers to shift risks and responsibilities on to workers" (p. 8).

12.4 CASES FOR AND AGAINST PLATFORM WORK FOR YOUNG PEOPLE

As already stated in this chapter, one of the pros of platform work is the flexibility it typically entails, as well as the provided low-threshold opportunities to work (Kuek et al., 2015). As Harmon and Silberman (2019) put it, platform work is a permanent part of modern-day work that is, by definition, flexible, digital, and globally networked. Those who promote platform work emphasize its ability to give workers autonomy and control within their working lives, while still partaking in the matchmaking between providers and customers in exchange for money (Lepanjuuri et al., 2018; Purcell & Brook, 2022). Some have also noted the ability to engage in various different forms of work without necessarily having any training or former experience (Kost et al., 2020). Another perceived benefit of platform work, in the sense of low-threshold accessibility, is that it provides workers who have limited skills (e.g., language skills or work experience) the opportunity to work, making it desirable to, for instance, migrant workers (Ly, 2020), or young people who are in need of part-time work or placeholder jobs while establishing their career trajectories (Kalleberg, 2009).

As well as platform work acting as a means to make a living, it also provides individuals the opportunity to make extra income in a flexible manner (e.g., students) (Ly, 2020; Sutela, 2023). Some opt to see platform work as an opportunity for innovations, with organizations utilizing digitalization to produce efficiency, stimulate consumption, and give individuals the ability to develop entrepreneurial skills (Dyal-Chand, 2015; Burtch et al., 2016). The flexibility, on the other hand, can be seen as beneficial for both workers and the companies that mediate these platform workers. While platform workers can determine their working hours, companies are relieved of their duties of having to schedule working hours (Hall & Krueger, 2018; Burtch et al., 2016). The additional benefits for these mediating companies include not having to be responsible for their employees in the same vein as traditional employers, but, instead, they can assign responsibility to their workers. These jobs and their accessibility can be seen as beneficial, particularly in times of financial or labour market turbulence, giving individuals options to make ends meet when other means of employment are less available (Purcell & Brook, 2022). Some scholars, however, argue that during economic crises, platform or gig work, in general, may be more of a liability than an advantage due to its precarious nature and lack of formality (Saad-Lessler, 2021). For young people, this can be seen as a double-edged sword, as during economic crises, young people are typically those who suffer from a lack of jobs and income stability (Ranta, 2015). For migrants, gig work is similarly considered simultaneously an opportunity and a risk (Van Doorn et al., 2023). On one hand, platform work can be one option to help weather the storm or obtain income regardless of language barriers or the difficulties of becoming employed; however, on the other, it typically does not promise either steady income or long-term career development, while also excluding workers from important welfare and social protection systems (Kalleberg, 2009; Laursen et al., 2021; Gerber, 2022; Van Doorn et al., 2023). Specifically in the case of migrant work, a criticism has emerged that gig work may be exploitative as migrants may lack the linguistic capabilities to defend their rights as workers while also being forced to partake in gig work as they otherwise struggle to find employment, perhaps even in expert positions more suited to their skills – primarily due to the fact that there is no incentive to aid them as they are, technically, not without salary (Van Doorn et al., 2023). Additionally, migrant workers run the risk of facing discrimination or potentially threatening encounters in their work due to biases (Anderson et al., 2021).

Many of the studies that have emerged in light of the expansion of gig work via platforms have solely focused on essential workers' rights or essential social protection systems due to a lack of regulations (Harmon & Silberman, 2019). In Europe, where trade unions and employer associations are heavily influential on the working culture and legalities of their countries, the debate about platform workers and their rights has been essential in work-related research and policymaking, as seen in reports by Eurofound and the European Trade Union Institute (Adam et al., 2023; Jepsen & Drahokoupil, 2017). In the United States, where trade unions and employer associations have less influence, there have still been discussions about what are the implications of growing platform work arrangements on increasing precarity in the labour force (Liebman, 2016). Due to platform workers often being labelled as self-employed, they are often deprived of many benefits and rights that are typical for employees – such as minimum wage, compensated sick leave, vacation days, parental leave, overtime pay, protection from discrimination, health insurance, and retirement – with little to no ability to negotiate their terms with the platforms that still rely on these workers (De Stefano, 2016; Liebman, 2016). Comprehensive research conducted by the Pew Research Center (Anderson et al., 2021) found that platform workers also reported that the risks associated with their work involved being treated rudely or harassed by customers and feeling unsafe. In addition to this, some stated that the customer review system which some platforms have for their gig workers has led to unwarranted poor reviews due to factors such as their race or other discriminatory reasons (Anderson et al., 2021).

Young people are forecast to become increasingly involved in platform work (Ilsøe & Larsen, 2020), which is assumed to also increase work-related vulnerabilities and risks (MacDonald & Giazitzoglu, 2019; Laursen et al., 2021). One of the postulated risks particularly involves work-related accidents, as it is believed that on-demand related work that involves carrying out numerous tasks throughout the day increases the likelihood of injuries and accidents during the workday (Garben, 2021; Laursen et al., 2021; Standing, 2018). These risks are particularly associated with young workers, as research on the topic has found that young people in particular attach minimal emphasis on risks when carrying out short-term employment (Garben, 2021; Laursen et al., 2021).

Some scholars have also criticized platform work for giving a false sense of autonomy (e.g., the flexibility of platform work) and for being falsely inviting to young people because of this. While there is much flexibility involved with platform work, there are also imperceptible limitations on autonomy, such as those regarding algorithmic management, non-transparency, and the platforms' alleged "invisible management" (Laursen et al., 2021; Gandini, 2019).

The low-threshold participation and flexibility of platform work is understandably tempting to young people, especially if alternative means of work are unavailable or if they are in need of part-time work only (Kalleberg, 2009; Hoang et al., 2020). This coincides with the findings from multiple countries that have found that young people are likely to engage in platform work, especially as a way to gain additional income (Hoang et al., 2020). However, some fear that platform work does not sufficiently provide young workers with work skills that would benefit them further along in their careers, given that very often platform work does not involve tasks that are either challenging or socially engaging (Purcell & Brook, 2022; Garben, 2021). This creates a risk opportunity in the sense that a young person may be tempted to engage with platform work as an easy solution, while in actuality they may benefit more from other, more traditional, part-time work, from a career development and networking perspective.

The main takeaway from the pros and cons of platform work is that, from a positive point of view, platform work is believed to reduce unemployment and cultivate flexible means to supplement income for low-income earners in particular (Graham et al., 2017; Jepsen & Drahokoupil, 2017; Hoang et al., 2020). However, from a negative perspective, it is believed that platform work creates disadvantages for workers, as employers may forgo essential protections that typical forms of employment provide for their workers. Furthermore, the precarious, typically low-paying, nature of platform work worsens inequalities among workers (Jepsen & Drahokoupil, 2017; Hoang et al., 2020; Lehdonvirta, 2018). In the context of young people, their typically favourable health, appreciation of flexibility, and common need for part-time work coincide very well with those who provide platform work and benefit from such workers (Woodcock, 2020). However, young people are also more vulnerable to financial precarity and typically have fewer monetary resources to overcome sudden, unexpected challenges that are a part of life. Since platform work does not provide the same safety nets as more traditional forms of employment, such as paid sick/vacation leave, parental leave, overtime pay, health insurance, pensions, compensation for work-related illnesses/injuries, and the right to safe working conditions, young workers run the risk of only truly benefiting from platform work until something goes awry (Hoang et al., 2020; Nielsen et al., 2022). Therein lies the question of whether the inequalities induced by platform work outweigh its benefits, as inequality is a constantly growing social problem in OECD countries (Hoang et al., 2020).

The COVID-19 pandemic served as a valuable example of how platform work is adjacent to the state of society and its economy. During the pandemic, food service platform work, in particular, became exceedingly popular as people were confined to their homes and some restaurants still wanted to run their businesses on a takeaway basis (Islam, 2022; Ravenelle et al., 2021; Wilska et al., 2021). Additionally, the furloughs and layoffs caused by the pandemic also introduced more people into the labour market who could supplement their losses via platform work (Apouey et al., 2020). In this example, the availability of platform work when other jobs are not as typically available is undoubtedly an asset, as many of the job losses and furloughs impacted young people specifically (Wilska et al., 2021). However, the pandemic also highlighted issues within platform work, such as the health inequalities it introduced (protective items such as masks and hand sanitizer were to be self-provided, and there was a lack of typical employer-provided health insurance). Additionally, those in platform work were negatively affected, such as those in transportation, since the demand for taxi services diminished during the pandemic (Ravenelle et al., 2021; Rani & Dhir, 2020). Thus, the pandemic serves as an ample example of both the flexibilities and opportunities provided by platform work, while also highlighting its tendencies to increase inequality and uncertainties (Apouey et al., 2020; Ravenelle et al., 2021; Rani & Dhir, 2020).

12.5 DISCUSSION

This chapter has reviewed the rise of informal employment and the prevalence of platform work in contemporary society and its implications for young people. It also briefly considers the implications for migrants in the interest of intersectionality. The advent of digital platforms has played a pivotal role in facilitating platform-based arrangements, creating a growing informal structure of employment (Watson et al., 2021; Woodcock & Graham, 2019). Platform work offers advantages, with flexibility being a central theme (Woodcock, 2020). However, this flexibility comes at a cost, as informal employment is characterized by unpredictability, job insecurity, and a lack of traditional benefits (Muntaner, 2018). The debate about whether platform work empowers individuals or perpetuates exploitation and inequalities remains a critical point of discussion (Woodcock & Graham, 2019; Hoang et al., 2020).

Sources have found that young people are inclined to partake in platform work, particularly on a part-time basis (Anderson et al., 2021; Sutela, 2023, ILO, 2021), with their participation forecast to increase (Ilsøe & Larsen, 2020). Similar trends are noted among non-whites and migrants (Anderson et al., 2021; Van Doorn et al., 2023). In this light, concerns about inequalities have emerged in discussions about platform work. For young people, these implications refer to the lack of opportunities to make meaningful connections, gain important work skills, and being put into a state of risk since platform work does not provide the same safety nets as traditional employment (Laursen et al., 2021; Gandini, 2019). For migrants, the caveats are largely the same, with further criticisms aimed at the potential exploitative nature of platform work and increased vulnerabilities regarding language barriers and the inability to find formal employment in one's host country (Van Doorn et al., 2023). While platform work provides low-threshold employment and flexible working hours, young age is considered a particularly high-risk time for individuals during critical career developmental stages, therefore raising the risk for long-term precarious employment (Arnett, 2014). Meanwhile, migrants may be disadvantaged by default in labour markets, leading to a situation where platform work is the only solution, stifling their opportunities for formal work and thus better safety nets and workers' rights (Van Doorn et al., 2023).

Advocates of platform work highlight its role in providing autonomy, flexibility, and opportunities for innovation. Platform work's ability to adapt to changing economic circumstances, especially during crises, is seen as a strength (Purcell & Brook, 2022), since platform companies place the economic responsibilities on their workers, while the workers can choose how they adapt to an economic crisis (e.g., taking more shifts) (Purcell & Brook, 2022). However, critics argue that gig work exacerbates precarity, as gig workers are often deprived of essential benefits and rights, leading to potential exploitation (De Stefano, 2016). Since young people are more likely to be negatively impacted by crises, including economically, there is the risk that general economic uncertainty will drive young people further into these forms of employment. The risk of this is the further exacerbation of financial and professional inequalities, which contribute to youth uncertainty (Arnett, 2014; Green, 2017). Particularly in light of recent economic turbulence, one could argue that there needs to be critical discussion about whether platform work can make young people more susceptible to being exploited, as they may become more dependent on platform work to make ends meet.

As an example, the pandemic brought both opportunities and challenges for platform workers. On one hand, it increased the demand for delivery services (Wilska et al., 2021; Apouey et al., 2020). On the other hand, the pandemic intensified existing vulnerabilities, exposing platform workers to health risks, insufficient safety measures, and financial instability (Apouey et al., 2020). Policies must strike a balance between preserving the flexibility that attracts individuals to platform work while ensuring fair treatment, social protections, and labour rights for platform workers. In conclusion, the dynamics of platform work are complex and multifaceted. While gig work offers a novel approach to employment and easy opportunities for young people, it also brings challenges that demand careful consideration, particularly in a cohort that is already considered particularly vulnerable and in an uncertain life stage (Arnett, 2014). The ongoing debates surrounding platform work highlight the need for a nuanced and adaptable approach to policymaking, while also conducting research that investigates the long-term implications of platform work on matters such as safety, health, and wealth accumulation. As society continues to navigate the evolving landscape of work, understanding the implications of platform work is crucial for shaping a future that is equitable, inclusive, and responsive to the diverse needs of the workforce.

12.5.1 Practical Implications

The practical implications of this chapter are to provide the reader with an overview of the modern, digital gig economy, as well as what the positive and negative aspects of its emergence provide to the realm of modern work. By taking into account the sociodemographic trends of those who partake in platform work, a clear trend can be seen in how young age and migrant/ethnic minority qualities are typically represented among platform workers. This invites further discussion as to whether the digital gig economy provides those in need with low-threshold employment options or exploits them.

12.5.2 Avenues for Future Research

Further research on the topic could explore what kind of careers develop in the gig economy and among those who have partaken in it. It could also examine the personal experiences and thoughts of those who partake in platform work regarding job satisfaction, security, and career prospects. Lastly, it would be particularly useful to explore the educational background of migrant gig workers to see if they are partaking in jobs below their qualifications due to stifled and/or difficult-to-access labour markets.

AUTHOR'S NOTE

This work was supported by the Strategic Research Council (SRC) at the Academy of Finland (#352544).

REFERENCES

Adam, M., Wiener, M., & Benlian, A. (2023). Online labor platforms and the role of job security and compensation (mis)fits for gig workers. 1–17. *ICIS 2023 Proceedings 2.*

Anderson, M., McClain, C., Faverio, M., & Gelles-Watnick, R. (2021). *The State of Gig Work in 2021*. Pew Research Center. https://www.pewresearch.org/internet/2021/12/08/the-state-of-gig-work-in-2021/.
Apouey, B., Roulet, A., Solal, I., & Stabile, M. (2020). Gig workers during the COVID-19 crisis in France: Financial precarity and mental well-being. *Journal of Urban Health*, *97*(6), 776–795. https://doi.org/10.1007/s11524-020-00480-4.
Arnett, J. J. (2014). Emerging adulthood: The winding road from the late teens throughthe twenties. Oxford University Press.
Arriagada, A., & Bishop, S. (2021). Between commerciality and authenticity: The imaginary of social media influencers in the platform economy. *Communication, Culture and Critique*, *14*(4), 568–586. https://doi.org/10.1093/ccc/tcab050.
Bosma, J. R. (2022). Platformed professionalization: Labor, assets, and earning a livelihood through Airbnb. *Environment and Planning A: Economy and Space*, *54*(4), 595–610. https://doi.org/10.1177/0308518X211063492.
Burtch, G., Carnahan, S., & Greenwood, B. N. (2016). Can you gig it? An empirical examination of the gig-economy and entrepreneurship. In *Academy of Management Proceedings* (Vol. 2016, No. 1, p. 14466). Briarcliff Manor, NY: Academy of Management. https://doi.org/10.5465/ambpp.2016.15.
Collier, R. B., Dubal, V., & Carter, C. (2017). *Labor Platforms and Gig Work: The Failure to Regulate*. IRLE Working Paper No. 106–17. http://irle.berkeley.edu/files/2017/Labor-Platforms-and-Gig-Work.pdf.
Davidson, A., Gleim, M. R., Johnson, C. M., & Stevens, J. L. (2023). Gig worker typology and research agenda: Advancing research for frontline service providers. *Journal of Service Theory and Practice*, *33*(5), 647–670. https://doi.org/10.1108/JSTP-08-2022-0188.
De Stefano, V. (2016). Introduction: Crowdsourcing, the gig-economy and the law. *Comparative Labor Law & Policy Journal*, *37*(3), 461–470. https://ssrn.com/abstract=2767383.
Dyal-Chand, R. (2015). Regulating sharing: The sharing economy as an alternative capitalist system. *Tulane Law Review*, *90*, 241–309. https://ssrn.com/abstract=2743959.
Florisson, R., & Mandl, I. (2018). *Platform Work: Types and Implications for Work and Employment – Literature Review*. Dublin: Eurofound. https://www.eurofound.europa.eu/system/files/2019-12/wpef18004.pdf.
Gandini, A. (2019). Labour process theory and the gig economy. *Human Relations*, *72*(6), 1039–1056. https://doi.org/10.1177/0018726718790002.
Garben, S. (2021). The regulation of platform work in the European Union: Mapping the challenges. In J. Drahokoupli & K. Vandaele (Eds), *A Modern Guide To Labour and the Platform Economy* (pp. 145–161). Cheltenham, UK and Northampton, MA, USA: Edward Elgar Publishing. https://doi.org/10.4337/9781788975100.00019.
Gerber, C. (2022). Gender and precarity in platform work: Old inequalities in the new world of work. *New Technology, Work and Employment*, *37*(2), 206–230. https://doi.org/10.1111/ntwe.12233.
Graham, M., Hjorth, I., & Lehdonvirta, V. (2017). Digital labour and development: Impacts of global digital labour platforms and the gig economy on worker livelihoods. *Transfer: European Review of Labour and Research*, *23*(2), 135–162. https://doi.org/10.1177/1024258916687250.
Green, A. (2017). *The Crisis for Young People: Generational Inequalities in Education, Work, Housing and Welfare*. Basingstoke: Palgrave Macmillan.

Griesbach, K., Reich, A., Elliott-Negri, L., & Milkman, R. (2019). Algorithmic control in platform food delivery work. *Socius, 5*. https://doi.org/10.1177/2378023119870041.
Hall, J. V., & Krueger, A. B. (2018). An analysis of the labor market for Uber's driver-partners in the United States. *ILR Review, 71*(3), 705–732. https://doi.org/10.1177/0019793917717222.
Harmon, E., & Silberman, M. S. (2019). Rating working conditions on digital labor platforms. *Computer Supported Cooperative Work (CSCW), 28*(5), 911–960. https://doi.org/10.1007/s10606-018-9313-5.
Hoang, L., Blank, G., & Quan-Haase, A. (2020). The winners and the losers of the platform economy: Who participates? *Information, Communication & Society, 23*(5), 681–700. https://doi.org/10.1080/1369118X.2020.1720771.
Huws, U., Spencer, N., Syrdal, D. S., & Holts, K. (2017). *Work in the European Gig Economy: Research Results from the UK, Sweden, Germany, Austria, the Netherlands, Switzerland and Italy*. Foundation for European Progressive Studies. https://uhra.herts.ac.uk/bitstream/handle/2299/19922/Huws_U._Spencer_N.H._Syrdal_D.S._Holt_K._2017_.pdf.
Ilsøe, A., & Larsen, T. P. (2020). Digital platforms at work: Champagne or cocktail of risks?. In *The impact of the sharing economy on business and society* (pp. 1-20). Routledge.
International Labour Organization. (2011). *From Precarious Work to Decent Work: Outcome Document to the Workers' Symposium on Policies and Regulations to Combat Precarious Employment*. International Labour Organization. https://www.ilo.org/wcmsp5/groups/public/---ed_dialogue/---actrav/documents/meetingdocument/wcms_179787.pdf.
International Labour Organization. (2021). *Young People and the Gig Economy*. International Labour Organization. https://www.ilo.org/wcmsp5/groups/public/---ed_emp/documents/publication/wcms_790117.pdf.
Islam, A. (2022). Work-from/at/for-home: COVID-19 and the future of work – A critical review. *Geoforum, 128*, 33–36. https://doi.org/10.1016/j.geoforum.2021.11.018.
Jepsen, M., & Drahokoupil, J. (2017). The digital economy and its implications for labour. 2. The consequences of digitalisation for the labour market. *Transfer: European Review of Labour and Research, 23*(3), 249–252. https://doi.org/10.1177/1024258917714659.
Kalleberg, A. L. (2009). Precarious work, insecure workers: Employment relations in transition. *American Sociological Review, 74*(1), 1–22. https://doi.org/10.1177/000312240907400101.
Kost, D., Fieseler, C., & Wong, S. I. (2020). Boundaryless careers in the gig economy: An oxymoron? *Human Resource Management Journal, 30*(1), 100–113. https://doi.org/10.1111/1748–8583.12265.
Kuek, S. C., Paradi-Guilford, C., Fayomi, T., Imaizumi, S., Ipeirotis, P., Pina, P., & Singh, M. (2015). *The Global Opportunity in Online Outsourcing*. Washington, DC: World Bank.
Laursen, C. S., Nielsen, M. L., & Dyreborg, J. (2021). Young workers on digital labor platforms: Uncovering the double autonomy paradox. *Nordic Journal of Working Life Studies, 11*(4), 65–84. https://doi.org/10.18291/njwls.127867.
Lehdonvirta, V. (2018). Flexibility in the gig economy: Managing time on three online piecework platforms. *New Technology, Work and Employment, 33*(1), 13–29. https://doi.org/10.1111/ntwe.12102.

Lepanjuuri, K., Wishart, R., & Cornick, P. (2018). *The Characteristics of Those in the Gig Economy.* London: BEIS.
Liebman, W. B. (2016). Labor law and the NLRB: Friend or foe to labor and non-union workers? *Harbinger, 41*(4), 4–10.
Lord, C., Bates, O., Friday, A., McLeod, F., Cherrett, T., Martinez-Sykora, A., & Oakey, A. (2023). The sustainability of the gig economy food delivery system (Deliveroo, UberEATS and Just-Eat): Histories and futures of rebound, lock-in and path dependency. *International Journal of Sustainable Transportation, 17*(5), 490–502. https://doi.org/10.1080/15568318.2022.2066583.
Ly, B. (2020). Employment in the gig economy. *On Research, 12*, 29–37. https://www.euruni.edu/Htdocs/Files/v/8575.pdf/downloads/foundationprograms/foundationprogramstrifold.pdf#page=29.
MacDonald, R., & Giazitzoglu, A. (2019). Youth, enterprise and precarity: Or, what is, and what is wrong with, the "gig economy"? *Journal of Sociology, 55*(4), 724–740. https://doi.org/10.1177/1440783319837604.
Mehta, B. S. (2023). Changing nature of work and the gig economy: Theory and debate. *FIIB Business Review, 12*(3), 227–237. https://doi.org/10.1177/2319714520968294.
Muntaner, C. (2018). Digital platforms, gig economy, precarious employment, and the invisible hand of social class. *International Journal of Health Services, 48*(4), 597–600. https://doi.org/10.1177/0020731418801413.
Nielsen, M. L., Laursen, C. S., & Dyreborg, J. (2022). Who takes care of safety and health among young workers? Responsibilization of OSH in the platform economy. *Safety Science, 149*, 105674. https://doi.org/10.1016/j.ssci.2022.105674.
Pfau-Effinger, B., Flaquer, L., & Jensen, P. H. (Eds). (2010). *Formal and Informal Work: The Hidden Work Regime in Europe* (Vol. 46). London: Routledge.
Prassl, J., & Risak, M. (2015). Uber, TaskRabbit, and co.: Platforms as employers: Rethinking the legal analysis of crowdwork. *Comparative Labour Law & Policy Journal, 37*, 619–648.
Purcell, C., & Brook, P. (2022). At least I'm my own boss! Explaining consent, coercion and resistance in platform work. *Work, Employment and Society, 36*(3), 391–406. https://doi.org/10.1177/0950017020952661
Rani, U., & Dhir, R. K. (2020). Platform work and the COVID-19 pandemic. *The Indian Journal of Labour Economics, 63*, 163–171. https://doi.org/10.1007/s41027-020-00273-y.
Ranta, M. (2015). Pathways to adulthood: Developmental tasks, financial resources and agency. Doctoral Dissertation, University of Jyväskylä. http://urn.fi/URN:ISBN:978-951-39-6227-2.
Ravenelle, A. J., Kowalski, K. C., & Janko, E. (2021). The side hustle safety net: Precarious workers and gig work during COVID-19. *Sociological Perspectives, 64*(5), 898–919. https://doi.org/10.1177/07311214211005489.
Ray, T. K., & Pana-Cryan, R. (2021). Work flexibility and work-related well-being. *International Journal of Environmental Research and Public Health, 18*(6), 3254. https://doi.org/10.3390/ijerph18063254.
Roy, G., & Shrivastava, A. K. (2020). Future of gig economy: Opportunities and challenges. *IMI Konnect, 9*(1), 14–27.
Saad-Lessler, J. (2021). *Pros and Cons of Gig Work During the COVID-19 Pandemic. Social Science Research Network.* Social Science Research Network. https://dx.doi.org/10.2139/ssrn.3877639.

Standing, G. (2018). Taskers in the precariat. In *Confronting Dystopia: The New Technological Revolution and the Future of Work.* Ithaca, NY: Cornell University Press.

Sutela, H. (2023). Alustatyötä vai ei? – Ilmiön rajaus ja nopea kehitys mutkistavat mittaamista. *Tilastokeskus.* https://www.stat.fi/tietotrendit/artikkelit/2023/alustatyota-vai-ei-ilmion-rajaus-ja-nopea-kehitys-mutkistavat-mittaamista/.

Vallas, S., & Schor, J. B. (2020). What do platforms do? Understanding the gig economy. *Annual Review of Sociology, 46*, 273–294. https://doi.org/10.1146/annurev-soc-121919–054857.

Van Doorn, N., Ferrari, F., & Graham, M. (2023). Migration and migrant labour in the gig economy: An intervention. *Work, Employment and Society, 37*(4), 1099–1111. https://doi.org/10.1177/09500170221096581.

Vyas, N. (2021). 'Gender inequality-now available on digital platform': an interplay between gender equality and the gig economy in the European Union. *European Labour Law Journal, 12*(1), 37–51.

Watson, G. P., Kistler, L. D., Graham, B. A., & Sinclair, R. R. (2021). Looking at the gig picture: Defining gig work and explaining profile differences in gig workers' job demands and resources. *Group & Organization Management, 46*(2), 327–361. https://doi.org/10.1177/1059601121996548.

Wilska, T., Sirola, A., Nuckols, J., Nyrhinen, J. (2021). *A Year of COVID-19 in Three Countries – A Study on the Effects of the COVID-19 Pandemic on Everyday Life, Consumption and Digital Behaviour in Finland, Sweden, and Great Britain.* YFI julkaisuja.9. Jyväskylän yliopisto. http://urn.fi/URN:ISBN:978-951-39-8829-6.

Woodcock, J. (2020). Technology, labor, and the gig economy. In D. A. Rohlinger & S. Sobieraj (Eds.), The Oxford handbook of digital media sociology (pp. 178–196). Oxford University Press.

Woodcock, J., & Graham, M. (2019). *The Gig Economy. A Critical Introduction.* Cambridge: Polity, 54.

PART III

Risks and opportunities for young people in digital environments

13. Problematic online behaviours during COVID-19: emerging risks and concerns

Anu Sirola

13.1 INTRODUCTION

The COVID-19 pandemic was a global crisis that began in early 2020, affecting the daily lives of people around the world. Countries and national regimes adopted varying strategies and lockdowns to prevent the spread of the virus, such as stay-at-home mandates, closures of many public institutions and establishments, and the prohibition of social gatherings (Wilska et al., 2021). Citizens were forced to change their daily lifestyles and spend time at home, physically isolated from others, becoming more dependent on the digital technologies that provided means for entertainment, social connection, and remote work or studying (Renu, 2021).

The pandemic-time safety measures and related 'digi leap' affected many consumption behaviours that have an addictive potential such as online shopping, gambling, and gaming (Maraz et al., 2021). Online shopping increased exponentially (Tyrväinen & Karjaluoto, 2022), gambling opportunities shifted to the online realm (Håkansson, 2020), and digital gaming was adopted as a way to spend time and socialize (Barr & Copeland-Stewart, 2022). Since the pandemic inflicted negative emotions such as distress and anxiety, some individuals coped with their negative emotional states by engaging in problematic and addictive online behaviours (Király et al., 2020; Maraz et al., 2021; Masaeli & Farhadi, 2021; Zamboni et al., 2021).

Young adults reported elevated feelings of psychological distress and loneliness in response to the COVID-19 crisis (Celuch et al., 2023; Glowacz & Schmits, 2020; Hawes et al., 2022; Shanahan et al., 2022). Young adulthood is a vulnerable life stage when identity is still developing, and important future decisions are made (Eriksson et al., 2020). The pandemic amplified the uncertainty of this stage, for example by increasing concerns regarding income and career (Nuckols et al., 2023). Adolescence and young adulthood are characterized by impulsivity and novelty seeking, making it a risky phase to develop problematic and potentially dangerous habits such as addictions (Chambers & Potenza, 2003; Petrović et al., 2022) and financial risk behaviour (Worthy et al., 2010). Young people also use online technologies and social media extensively, thus being at risk of encountering various dangers in online environments (Ferrara et al., 2021; Purba et al., 2023). Therefore, young people were vulnerable to various problem behaviours and consequent risks during the pandemic.

This chapter synthesizes research evidence regarding young people's risky online consumption behaviours during the COVID-19 pandemic, focusing on addictive behaviours and consumption of risky information sources such as the dark web. The approach of this chapter is twofold. It examines general risks related to online platforms and their potential to shape one's attitudes and behaviours. Additionally, the chapter examines problematic online gambling, gaming, and online shopping during the pandemic. Such activities are treated as examples of risky consumption behaviours facilitated by online environments. Gambling and shopping are inherently monetary activities, posing risks to financial well-being (Nyrhinen et al., 2023; Oksanen et al., 2018). In addition, digital gaming has monetary features such as loot boxes (e.g., crates that contain game-related items) that are a form of risky and gambling-like consumption (Brooks & Clark, 2019).

This chapter is organized as follows. First, pandemic-time online behaviours are examined from a social psychological perspective, looking at the role of online technologies in social interaction, information seeking, and social influence. After that, the chapter examines problematic online consumption behaviours during the pandemic. Finally, the key findings and long-lasting effects of the pandemic are discussed, also providing implications for research and practice. The focus is on risks, but also the benefits of online technologies are discussed in relation to crisis situations.

13.2 ONLINE TECHNOLOGIES AS MEANS FOR INTERACTION, INFORMATION, AND INFLUENCE

13.2.1 Social Aspects of Online Technologies

The first years of the COVID-19 pandemic (2020–2022) were characterized by physical distancing and excessive social isolation as safety measures to prevent the spread of the virus (Wilska et al., 2021), posing a threat to the basic human need for social connection (Baumeister & Leary, 1995; Deci & Ryan, 2000). While loneliness did not significantly increase in general populations, social isolation had the most detrimental impact on vulnerable individuals such as young people, among whom isolation elevated experiences of loneliness (Lee et al., 2020; Luchetti et al., 2020). Loneliness has major clinical significance, as it is associated with various deficits in both mental and physical well-being (Hawkley & Cacioppo, 2010; Heinrich & Gullone, 2006; Latikka et al., 2022).

Loneliness motivates individuals to seek connections in the online realm, particularly among young adults who use such technologies extensively (O'Day & Heimberg, 2021; Sirola et al., 2019). During the pandemic and considering social isolation, online networks provided meaningful social resources to cope with distressing situations but did not protect against psychological distress among individuals who often felt themselves lonely (Latikka et al., 2022). At their best, online interactional tools can support already existing relationships but do not fully compensate for a lack of meaningful connections (Nowland et al., 2018). Moreover, excessive internet and social media use might further reinforce feelings of loneliness (Moretta & Buodo, 2020).

Social media facilitates the seeking of like-minded others and the forming of online communities around shared interests. Online communities might manifest themselves in online identity bubbles that are characterized by identification with like-minded social networks and reliance on shared information (Kaakinen et al., 2020). The risks of online bubbles lie in one-sided exposure to misleading or harmful information, such as anti-vaccine movements or conspiracy theories.

Social norms are inherently present on social media. For example, other users' reactions and comments to content are prone to influence one's attitudes and behaviours, providing information on what kind of content is acceptable to share and like (Kim et al., 2015; Sirola et al., 2021a). Perceived social norms online have the potential to shape and influence potentially addictive behaviours such as gambling (Savolainen et al., 2021; Sirola et al., 2021a), gaming (Angelini et al., 2024), and online shopping (Lee et al., 2011). Young people also utilize various social media platforms as information sources, being susceptible to the social influence of their peers' opinions (Bene, 2017). This is a particular risk when shared information is of compromised credibility.

13.2.2 Risky Online Information Sources: the Case of the Dark Web

Social media platforms are extensively utilized as information sources, especially among younger age groups (Bene, 2017; Wilska et al., 2021). Social media contains diverse sets of reliable, unreliable, and even dangerous information from different sources. False information is likely to spread more quickly and be more popular compared with verified information (Vosoughi et al., 2018; Wang et al., 2019). During the pandemic, information regarding the virus, including both reliable and unreliable information, spread rapidly on various online platforms (Cinelli et al., 2020).

During the pandemic, concerns were raised regarding young people's increased interest in the dark web (Ferrara et al., 2021). The dark web is a hidden part of the internet not accessible with common browsers or search engines, mostly accessed via a TOR network (Montasari & Boon, 2023). Layered encrypting technologies facilitate high anonymity, which provides benefits for individuals who wish to browse the web anonymously, but it also encourages the propagation and consumption of illicit content and services, including marketplaces for illicit drugs (ElBahrawy et al., 2020). Because the content on the dark web is not regulated, the platforms allowed the propagation and flourishing of false information and conspiracy theories regarding the COVID-19 pandemic (Topor & Shuker, 2020). Moreover, forged COVID certificates were sold in dark web marketplaces as a means to bypass quarantines and other safety restrictions set by institutional authorities to prevent the spread of the virus (Bracci et al., 2021; Broadhurst et al., 2020).

A study by Sirola et al. (2022) found that COVID-19 scepticism, loneliness, and high engagement in online activities, such as social media and online gambling, were associated with dark web use as a COVID-19 information source. Young adults and males were more likely users of the dark web. The results indicated that the dark web attracts advanced online users who have underlying psychosocial issues and are suspicious of official authorities. Even though only a minority reported having utilized the dark web as a COVID-19 information source, they are likely to be active online users who might spread such information to public social media, thus gaining visibility. Active online users might also be more susceptible to encountering illicit content on the dark web, such as illegal gambling sites and marketplaces for illicit purchases.

13.3 INTERNET-BASED ADDICTIVE BEHAVIOURS DURING THE PANDEMIC: EXCESSIVE GAMBLING, GAMING, AND ONLINE SHOPPING

The COVID-19 pandemic changed the lifestyles of individuals and caused significant distress, which for some increased the potential to develop addictive behaviours (Alimoradi et al., 2022). Both internet use and internet-based addictive behaviours increased during the pandemic, with younger age groups being the most vulnerable to such issues (Maraz et al., 2021; Masaeli & Farhadi, 2021; Petrović et al., 2022). The increase was most notable in excessive online gaming and shopping (Maraz et al., 2021). Such behaviours might have been used to alleviate negative feelings such as anxiety or stress and were used as maladaptive coping strategies during distressing situations (Maraz et al., 2021).

Next, we will take a closer look at the research evidence on problematic online gambling, gaming, and shopping during the pandemic. These behaviours have addictive potential that inflicts emotional, social, and financial harm when developed to excess levels (Grant et al., 2010). The term 'excessive' is used as an umbrella term to cover potentially problematic and addictive forms of these behaviours. The term derives from Jim Orford's (2001) book *Excessive Appetites* in which he treats addictions as an extreme form of appetite where an individual becomes strongly attached to a certain activity, resulting in diminished self-control and emotional and social costs. Thus, the focus is not on the clinical diagnostic criteria of addictions, but a more general approach to harm-inflicting problematic behaviours facilitated by the internet.

13.3.1 Problematic Online Gambling

During the first years of the pandemic, gambling opportunities were significantly decreased due to the closure of physical gambling venues and the cancellation of major sports events. Therefore, the role of online gambling became greater as online gambling sites, such as online casinos, were accessible 24/7 (Emond et al., 2022; Håkansson, 2020). Prior to the pandemic, the estimated prevalence rates of problem gambling among adolescents and young people worldwide varied from 0.2 to 12.3 per cent (Calado et al., 2017). Concerns were raised that online gambling and related problems would increase as a response to the pandemic and isolation (Brodeur et al., 2021).

According to evidence, gambling activities or problems did not increase on a population level; however, problems elevated among some individuals, such as those who had excessive gambling habits prior to the pandemic (Brodeur et al., 2021; Emond et al., 2022; Hodgins & Stevens, 2021; Håkansson, 2020). A review study by Masaeli and Farhadi (2021) concluded that the pandemic-time increase in online gambling was correlated with anxiety, depression, problematic substance use, and financial problems. Emond et al. (2022) found that heavy alcohol use and financial difficulties prior to the pandemic were risk factors for regular engagement in gambling during COVID-19 lockdowns in the UK.

A study by Sirola et al. (2023a) surveyed 18–75-year-old respondents from Finland, Sweden, and the UK, looking at the psychosocial risk and protective factors for pandemic-time problem gambling. The results revealed that younger age and pandemic-time loneliness were significant risk factors for problem gambling. Interestingly, perceived social support did not protect from problem gambling but was associated with greater problem gambling severity. Even though social support is generally perceived as a protective factor against many risks, support derived from online gambling communities that are based on positive views of gambling can facilitate excessive gambling (Savolainen et al., 2022a; Sirola et al., 2021b). Similarly, Savolainen et al. (2022b) found that social motives were associated with greater problem gambling severity during the COVID-19 pandemic. Given the isolated nature of online gambling, it is likely that such social motives are at least partially related to online gambling communities.

13.3.2 Problematic Online Gaming and Loot Box Purchasing

Digital gaming (i.e., online gaming) refers to games played with consoles, mobile phones, or computers. Different from online gambling, which is predominantly an isolated activity, social interaction is often inherently embedded in digital games as players collaborate in teams or guilds and interact with each other during the game (Sirola et al., 2021b). In the early stage of the COVID-19 pandemic, the World Health Organization (WHO) launched a #PlayApartTogether initiative to promote online gaming as a recreational activity and a safe way to connect with others in physical isolation (King et al., 2020).

The prevalence of online gaming increased during COVID-19 (López-Cabarcos et al., 2020; Maraz et al., 2021, Masaeli & Farhadi, 2021), the increase being most dramatic at key points of the pandemic (Vuorre et al., 2021). Globally, the time spent in digital gaming increased by 39 per cent during the first year of the pandemic (Clement, 2021). The popularity of multiplayer games increased more drastically compared with single-player games, emphasizing the social aspects of these games during isolation (Barr & Copeland-Stewart, 2022; Vuorre et al., 2021). Gaming for social purposes helped some individuals cope with pandemic-related emotional distress and social isolation and satisfy their psychological needs such as competence, autonomy, and relatedness (Yee & Sng, 2022). The pandemic also significantly increased the popularity of and time spent on Twitch, a streaming platform mostly known for gaming content (Clement, 2023).

Even though gaming helped to cope with pandemic-time distress and isolation, some individuals were vulnerable to developing unhealthy gaming habits (King et al., 2020; Király et al., 2020; Masaeli & Farhadi, 2021). Among young adults, psychological distress and using online gaming to cope with negative emotions were associated with increased and problematic gaming (Angelini et al., 2024; Claesdotter-Knutsson et al., 2022). Moreover, despite the mental health benefits of social opportunities in digital games, the social aspect of digital games also poses risks regarding problematic gaming. Among young people, social norms related to gaming as well as peers' perceived gaming frequency were associated with problematic gaming (Angelini et al., 2024). Moreover, social motives were found to be a risk factor for gaming problems during the pandemic (Savolainen et al., 2022b). Even though social interaction is one of the key motives in digital games, the social aspect of the games might make individuals susceptible to excessive gaming because of peer pressure to invest in the team's game success (Sirola et al., 2021b). This highlights the controversial nature of social interaction within digital games.

Digital games increasingly contain monetary and gambling-like features to enhance the gaming experience. The loot box system is an example of a gambling-like random-reward monetary element in digital games that has addictive potential (Brooks & Clark, 2019). Loot boxes contain a randomized set of items such as weapons, equipment, or cosmetic features. They are often purchasable with real money, with prices varying from a few to tens of dollars (Close et al., 2021). The player does not know the content of the loot box prior to purchasing, a feature that has striking similarities with gambling. There is a growing body of research showing that active loot box purchasing and excessive gambling co-occur, even though the causal mechanisms between these behaviours are still unclear (Brooks & Clark, 2019; Spicer et al., 2022). Loot boxes do not inflict harm on every player (Etchells et al., 2022). However, a minority of players spend over a hundred dollars or more per month on loot boxes (Close et al., 2021).

During the pandemic, contamination concerns were associated with greater loot box purchasing (Hall et al., 2021). Additionally, socially isolated gamers who had co-occurring problem gambling symptomatology were more at risk of loot box purchasing (Hall et al., 2021). Among adolescents, there was an increase in loot box spending during the COVID-19 lockdowns (DeCamp & Daly, 2023). Thus, crisis situations might facilitate increased loot box consumption among vulnerable individuals, which can accumulate financial strain and lead to further distress.

A study by Sirola et al. (2023b) examined loot box behaviour and associated risk and protective factors during COVID-19, showing that pandemic-time loneliness was associated with increased loot box activity. Loot box purchasing was also associated with problem gambling, which was further linked to indebtedness. The key message was that even though loot boxes may not in themselves strain one's economic situation, they might contribute to worsening economic strain, particularly among individuals who are in a vulnerable position, such as young people and problem gamblers.

13.3.3 Problematic Online Shopping

During the pandemic, there was an exponential growth in online shopping (Tyrväinen & Karjaluoto, 2022). According to *The New York Times*, the leading e-commerce market, Amazon, increased its profit by 220 per cent during the pandemic (Weise, 2021). For some individuals, online shopping triggers addictive and compulsive tendencies and develops into excess. During the early onset of the pandemic, problematic online shopping increased (Maraz et al., 2021). Problematic and compulsive forms of online shopping (i.e., online shopping addiction) share many commonalities with other behavioural addictions, such as mood modification, tolerance, and withdrawal symptoms (Rose & Dhandayudham, 2014; Zhao et al., 2017). Such purchase behaviours also have the potential to damage one's economic well-being (Nyrhinen et al., 2023).

A study by Nyrhinen et al. (2023) utilized pandemic-time survey data from Finnish 18- to 29-year-olds, examining young adults' online shopping addiction and economic difficulties. The study found that low self-regulation in an online environment is a risk factor for online shopping addiction and further leads to dissatisfaction with personal money management through indebtedness. Moreover, those young adults who had problems regulating their smartphone use and had low self-regulation skills in general were vulnerable to developing problematic online shopping habits.

Other pandemic-time research on problematic online shopping reveals that social isolation, anxiety, boredom, and extensive free time were risk factors for shopping addiction (Khrouf & Maghraoui, 2023; Niedermoser et al., 2021). Similarly, Zamboni et al. (2021) found that the loss of control in internet use and online shopping was associated with negative emotional states such as anxiety and nervousness. A study by Wang et al. (2022) found that among college students, academic procrastination and negative emotions were associated with online shopping addiction during the pandemic. These findings indicate that online shopping may serve as a maladaptive way to alleviate one's negative emotions during the crisis. However, online shopping may cause even more negative emotions and financial loss (Wang et al., 2022). Again, problematic shopping habits seem to overlap with other behavioural addictions such as problematic gambling and gaming (Puiras et al., 2022), emphasizing that various risk behaviours are likely to accumulate in online environments.

13.4 DISCUSSION

This chapter has synthesized research evidence on young adults' problematic online behaviours during the COVID-19 pandemic and consequent risks. The key finding of this chapter is that young adults were a highly vulnerable group for developing pandemic-time problem behaviours. Even though excessive online behaviours did not increase at a population level, problems were amplified among vulnerable individuals such as young people. Psychological distress and loneliness were major risks in adopting and reinforcing such maladaptive behavioural patterns, and young adults reported significantly high experiences of these issues (Celuch et al., 2023; Glowacz & Schmits, 2020; Hawes et al., 2022; Latikka et al., 2022; Shanahan et al., 2022).

This chapter has highlighted the twofold nature of digital environments in facilitating both well-being and risk behaviours. On one hand, digital technologies such as social media and digital games provide means to fulfil a basic psychological need to belong (Baumeister & Leary, 1995; Deci & Ryan, 2000). During the COVID-19 pandemic and physical isolation, the social aspects of these technologies were particularly beneficial and helped people to cope with loneliness and boredom. On the other hand, online behaviours can easily develop into excess and even addiction, leading to various deficits in well-being, particularly in vulnerable life situations. Addictions are prone to damage one's social, emotional, and financial well-being, and many problems can accumulate over the course of addictive behaviours. For example, financial insecurity and the wish to earn money can motivate gambling, but excessive gambling is likely to add to one's financial strain and cause further distress (Oksanen et al., 2018).

The findings also illustrate that problematic online behaviours tend to overlap, and various risks are likely to accumulate in online environments. Young people are the most active age group regarding the use of various online technologies and social media sites, including risky online platforms such as the dark web, making them susceptible to various risks and harmful peer influence (Savolainen et al., 2021; Sirola et al., 2021a; Tuominen et al., 2023). Digital environments provide constant distractive stimuli such as targeted advertising, which makes users susceptible to unplanned and impulsive purchase behaviours (Nyrhinen et al., 2023). Additionally, online gambling and gaming are fast-paced and addictive by nature, making it easy to get immersed in these activities. Thus, even though young people are active agents online and skilled technology users in general, online environments facilitate many risky behaviours and addictive tendencies, and the pandemic-time likely amplified the risks.

The key findings and arguments of this chapter are in line with the Interaction of Person-Affect-Cognition-Execution model (I-PACE; Brand et al., 2016) that provides a theoretical framework for the development and maintenance of addictive online behaviours such as gambling, gaming, and shopping. The I-PACE model considers the development of addictive behaviours as a process where personal, emotional, social, and situational factors are in constant interplay. According to the model, engaging in online behaviours can be a way to cope with negative emotional states, but these behaviours might worsen the distress and cause more problems, leading to a vicious cycle and addictive tendencies (Brand et al., 2016). In this chapter, the COVID-19 pandemic provided a meaningful situational context to examine how crisis situations can amplify online risky behaviours among vulnerable individuals, such as young people, helping us to understand the underlying socio-psychological mechanisms and risk factors that might contribute to such behaviours.

13.4.1 Practical Implications

In moderate forms, all online behaviours examined in this chapter may have been beneficial during the pandemic and associated social isolation, supporting basic psychological needs such as social belonging (Deci & Ryan, 2000; Yee & Sng, 2022). In the digitalized world, it is a great asset to know how to use different online technologies for social interaction, leisure activities, and information seeking, as online technologies can support a healthy lifestyle and social connections. For the majority of individuals, online gambling, gaming, or shopping do not develop into excessive behaviour or cause significant harm. However, given that some individuals are more vulnerable to develop unhealthy behavioural patterns (King et al., 2020), underlying vulnerabilities should be better recognized.

Throughout this chapter, two risk factors have been recognized for various problematic online behaviours during the pandemic: loneliness and psychological distress. In the clinical context, it is crucial to recognize these vulnerabilities and how they can increase the risk for various problematic behaviours. It would have been highly important to support young people's mental health and meaningful social connections during the crisis to avoid consequent risks. Young adulthood is a vulnerable time with regard to addictions and the accumulation of related problems, partly due to elevated levels of impulsivity and novelty seeking (Chambers & Potenza, 2003). Moreover, in young adulthood, the financial situation is often unstable, and excessive consumption behaviours may damage one's financial well-being in a long-lasting manner (Oksanen et al., 2018). Therefore, during crises, mental health interventions should be targeted at the most vulnerable groups such as young people.

The pandemic period made visible the many risks of unreliable, and even dangerous, information being shared on social media and the dark web (Cinelli et al., 2020; Sirola et al., 2022). Since young people prefer online platforms as an information source, it is crucial to understand where and why individuals seek information, particularly during crises, why risky platforms such as the dark web attract some users, and how perceived social norms online can further shape one's attitudes and behaviours in a harmful way. In online environments, false information spreads quickly, and algorithms further play a role in increasing the visibility of certain content.

To mitigate the risks of online environments, the focus should be on how to utilize online platforms in a more beneficial manner to support well-being and facilitate meaningful social connections. Regarding information abundance and potentially risky information sources, it is important to improve the media literacy of young people and to educate them on the dangers of algorithms in increasing the visibility of potentially harmful and one-sided content.

13.4.2 Avenues for Future Research

At the time of writing, in the spring of 2024, the world has returned to a (new) normal after the pandemic. However, in the aftermath, new crises have emerged, such as Russia's military attack on Ukraine, starting in early 2022, which is the biggest military conflict in Europe since the Second World War. Additionally, the ongoing energy crisis and growing inflation rate have created uncertainty and distress regarding the future (Nuckols et al., 2024). Some individuals are susceptible to the accumulation of negative mental health effects and consequent risks related to ongoing crises (Celuch et al., 2023). During the pandemic, people relied on online technologies, and many engaged in potentially addictive online behaviours such as gaming, gambling, and shopping as a coping mechanism to alleviate boredom, isolation, and distress. The risk is that these behaviours persist after the pandemic and are utilized as maladaptive coping mechanisms during other crises as well. Therefore, it is crucial to investigate how both the pandemic and the current crises affect problematic behaviours and mental health, and how these effects might accumulate over time.

AUTHOR'S NOTE

This work was supported by the Strategic Research Council (SRC) at the Academy of Finland (#352544).

REFERENCES

Alimoradi, Z., Lotfi, A., Lin, C. Y., Griffiths, M. D., & Pakpour, A. H. (2022). Estimation of behavioral addiction prevalence during COVID-19 pandemic: A systematic review and meta-analysis. *Current Addiction Reports*, *9*(4), 486–517.

Angelini, F., Pivetta, E., Marino, C., Canale, N., Spada, M. M., & Vieno, A. (2024). Social norms and problematic gaming among adolescents: The role of Internet use coping motives. *Addictive Behaviors*, *148*, 107865.

Barr, M., & Copeland-Stewart, A. (2022). Playing video games during the COVID-19 pandemic and effects on players' well-being. *Games and Culture*, *17*(1), 122–139.

Baumeister, R. F., & Leary, M. R. (1995). The need to belong: Desire for interpersonal attachments as a fundamental human motivation. *Psychological Bulletin*, *117*(3), 497–529.

Bene, M. (2017). Influenced by peers: Facebook as an information source for young people. *Social Media+Society*, *3*(2), 2056305117716273.

Bracci, A., Nadini, M., Aliapoulios, M., McCoy, D., Gray, I., Teytelboym, A., … & Baronchelli, A. (2021). Dark web marketplaces and COVID-19: Before the vaccine. *EPJ Data Science*, *10*(1), 6.

Brand, M., Young, K. S., Laier, C., Wölfling, K., & Potenza, M. N. (2016). Integrating psychological and neurobiological considerations regarding the development and maintenance of specific Internet-use disorders: An Interaction of Person-Affect-Cognition-Execution (I-PACE) model. *Neuroscience & Biobehavioral Reviews*, *71*, 252–266.

Broadhurst, R., Ball, M., & Jiang, C. J. (2020). Availability of COVID-19 related products on Tor darknet markets. *Australasian Policing*, *12*(3), 8–13.

Brodeur, M., Audette-Chapdelaine, S., Savard, A. C., & Kairouz, S. (2021). Gambling and the COVID-19 pandemic: A scoping review. *Progress in Neuro-Psychopharmacology and Biological Psychiatry*, *111*, 110389.

Brooks, G. A., & Clark, L. (2019). Associations between loot box use, problematic gaming and gambling, and gambling-related cognitions. *Addictive Behaviors*, *96*, 26–34.

Calado, F., Alexandre, J., & Griffiths, M. D. (2017). Prevalence of adolescent problem gambling: A systematic review of recent research. *Journal of Gambling Studies*, *33*, 397–424.

Celuch, M., Savolainen, I., Oksa, R., & Oksanen, A. (2023). Psychological distress and perceived stress outcomes of anxieties related to COVID-19 and Russia's attack on Ukraine. *Journal of Loss and Trauma*, 1–26.

Chambers, R. A., & Potenza, M. N. (2003). Neurodevelopment, impulsivity, and adolescent gambling. *Journal of Gambling Studies*, *19*, 53–84.

Cinelli, M., Quattrociocchi, W., Galeazzi, A., Valensise, C. M., Brugnoli, E., Schmidt, A. L., … & Scala, A. (2020). The COVID-19 social media infodemic. *Scientific Reports*, *10*(1), 1–10.

Claesdotter-Knutsson, E., André, F., & Håkansson, A. (2022). Gaming activity and possible changes in gaming behavior among young people during the COVID-19 pandemic: Cross-sectional online survey study. *JMIR Serious Games*, *10*(1), e33059.

Clement, J. (2021). Increase in time spent video gaming during the COVID-19 pandemic worldwide as of June 2020, by region. Statista Research Reports. https://www.statista.com/statistics/1188545/gaming-time-spent-covid/

Clement, J. (2023). Number of minutes spent watching Twitch content worldwide 2012–2023. Statista Research Reports. https://www.statista.com/statistics/819967/time-spent-watching-twitch/.

Close, J., Spicer, S. G., Nicklin, L. L., Uther, M., Lloyd, J., & Lloyd, H. (2021). Secondary analysis of loot box data: Are high-spending 'whales' wealthy gamers or problem gamblers? *Addictive Behaviors*, *117*, 106851.

DeCamp, W., & Daly, K. (2023). Loot box consumption by adolescents pre-and post-pandemic lockdown. *PeerJ*, *11*, e15287.

Deci, E. L., & Ryan, R. M. (2000). The 'what' and 'why' of goal pursuits: Human needs and the self-determination of behavior. *Psychological Inquiry*, *11*(4), 227–268.

ElBahrawy, A., Alessandretti, L., Rusnac, L., Goldsmith, D., Teytelboym, A., & Baronchelli, A. (2020). Collective dynamics of dark web marketplaces. *Scientific Reports*, *10*(1), 18827.

Emond, A., Nairn, A., Collard, S., & Hollén, L. (2022). Gambling by young adults in the UK during COVID-19 lockdown. *Journal of Gambling Studies*, *38*(1), 1–13.

Eriksson, P. L., Wängqvist, M., Carlsson, J., & Frisén, A. (2020). Identity development in early adulthood. *Developmental Psychology*, *56*(10), 1968–1983.

Etchells, P. J., Morgan, A. L., & Quintana, D. S. (2022). Loot box spending is associated with problem gambling but not mental wellbeing. *Royal Society Open Science*, *9*(8), 220111.

Ferrara, P., Franceschini, G., Corsello, G., Mestrovic, J., Giardino, I., Vural, M., … & Pettoello-Mantovani, M. (2021). The dark side of the web: A risk for children and

adolescents challenged by isolation during the novel coronavirus 2019 pandemic. *The Journal of Pediatrics, 228*, 324–325.
Glowacz, F., & Schmits, E. (2020). Psychological distress during the COVID-19 lockdown: The young adults most at risk. *Psychiatry Research, 293*, 113486.
Grant, J. E., Potenza, M. N., Weinstein, A., & Gorelick, D. A. (2010). Introduction to behavioral addictions. *The American Journal of Drug and Alcohol Abuse, 36*(5), 233–241.
Hall, L. C., Drummond, A., Sauer, J. D., & Ferguson, C. J. (2021). Effects of self-isolation and quarantine on loot box spending and excessive gaming: Results of a natural experiment. *PeerJ, 9*, e10705.
Hawes, M. T., Szenczy, A. K., Klein, D. N., Hajcak, G., & Nelson, B. D. (2022). Increases in depression and anxiety symptoms in adolescents and young adults during the COVID-19 pandemic. *Psychological Medicine, 52*(14), 3222–3230.
Hawkley, L. C., & Cacioppo, J. T. (2010). Loneliness matters: A theoretical and empirical review of consequences and mechanisms. *Annals of Behavioral Medicine, 40*(2), 218–227.
Heinrich, L. M., & Gullone, E. (2006). The clinical significance of loneliness: A literature review. *Clinical Psychology Review, 26*(6), 695–718.
Hodgins, D. C., & Stevens, R. M. (2021). The impact of COVID-19 on gambling and gambling disorder: Emerging data. *Current Opinion in Psychiatry, 34*(4), 332–343.
Håkansson, A. (2020). Changes in gambling behavior during the COVID-19 pandemic: A web survey study in Sweden. *International Journal of Environmental Research and Public Health, 17*(11), 4013.
Kaakinen, M., Sirola, A., Savolainen, I., & Oksanen, A. (2020). Shared identity and shared information in social media: Development and validation of the identity bubble reinforcement scale. *Media Psychology, 23*(1), 25–51.
Khrouf, L., & Maghraoui, S. (2023). Online shopping spree in times of COVID-19 pandemic. *International Journal of Electronic Business, 18*(2), 132–147.
Kim, S., Lee, J., & Yoon, D. (2015). Norms in social media: The application of theory of reasoned action and personal norms in predicting interactions with Facebook page like ads. *Communication Research Reports, 32*(4), 322–331.
King, D. L., Delfabbro, P. H., Billieux, J., & Potenza, M. N. (2020). Problematic online gaming and the COVID-19 pandemic. *Journal of Behavioral Addictions, 9*(2), 184–186.
Király, O., Potenza, M. N., Stein, D. J., King, D. L., Hodgins, D. C., Saunders, J. B., … & Demetrovics, Z. (2020). Preventing problematic internet use during the COVID-19 pandemic: Consensus guidance. *Comprehensive Psychiatry, 100*, 152180.
Latikka, R., Koivula, A., Oksa, R., Savela, N., & Oksanen, A. (2022). Loneliness and psychological distress before and during the COVID-19 pandemic: Relationships with social media identity bubbles. *Social Science & Medicine, 293*, 114674.
Lee, C. M., Cadigan, J. M., & Rhew, I. C. (2020). Increases in loneliness among young adults during the COVID-19 pandemic and association with increases in mental health problems. *Journal of Adolescent Health, 67*(5), 714–717.
Lee, M. K., Shi, N., Cheung, C. M., Lim, K. H., & Sia, C. L. (2011). Consumer's decision to shop online: The moderating role of positive informational social influence. *Information & Management, 48*(6), 185–191.
López-Cabarcos, M. Á., Ribeiro-Soriano, D., & Piñeiro-Chousa, J. (2020). All that glitters is not gold. The rise of gaming in the COVID-19 pandemic. *Journal of Innovation & Knowledge, 5*(4), 289–296.

Luchetti, M., Lee, J. H., Aschwanden, D., Sesker, A., Strickhouser, J. E., Terracciano, A., & Sutin, A. R. (2020). The trajectory of loneliness in response to COVID-19. *American Psychologist, 75*(7), 897–908.

Maraz, A., Katzinger, E., & Yi, S. (2021). Potentially addictive behaviours increase during the first six months of the Covid-19 pandemic. *Journal of Behavioral Addictions, 10*(4), 912–919.

Masaeli, N., & Farhadi, H. (2021). Prevalence of Internet-based addictive behaviors during COVID-19 pandemic: A systematic review. *Journal of Addictive Diseases, 39*(4), 468–488.

Montasari, R., & Boon, A. (2023, January). An analysis of the dark web challenges to digital policing. In *Cybersecurity in the Age of Smart Societies: Proceedings of the 14th International Conference on Global Security, Safety and Sustainability*, London, September 2022 (pp. 371–383). Cham: Springer International Publishing.

Moretta, T., & Buodo, G. (2020). Problematic Internet use and loneliness: How complex is the relationship? A short literature review. *Current Addiction Reports, 7*, 125–136.

Niedermoser, D. W., Petitjean, S., Schweinfurth, N., Wirz, L., Ankli, V., Schilling, H., … & Walter, M. (2021). Shopping addiction: A brief review. *Practice Innovations, 6*(3), 199–207.

Nowland, R., Necka, E. A., & Cacioppo, J. T. (2018). Loneliness and social internet use: Pathways to reconnection in a digital world? *Perspectives on Psychological Science, 13*(1), 70–87.

Nuckols, J. A., Silinskas, G., Ranta, M., & Wilska, T. A. (2023). Income and career concerns among emerging adults from Finland, Sweden, and the United Kingdom during COVID-19. *Emerging Adulthood, 11*(3), 721–734.

Nuckols, J., Wilska, T. A., & Sirola, A. (2024). A year after COVID-19 in three countries: The impact of the war in Ukraine and economic crisis on everyday life and wellbeing in Finland, Sweden, and the UK. *YFI julkaisuja*, (19).

Nyrhinen, J., Lonka, K., Sirola, A., Ranta, M., & Wilska, T. A. (2023). Young adults' online shopping addiction: The role of self-regulation and smartphone use. *International Journal of Consumer Studies*. https://doi.org/10.1111/ijcs.12961.

O'Day, E. B., & Heimberg, R. G. (2021). Social media use, social anxiety, and loneliness: A systematic review. *Computers in Human Behavior Reports, 3*, 100070.

Oksanen, A., Savolainen, I., Sirola, A., & Kaakinen, M. (2018). Problem gambling and psychological distress: A cross-national perspective on the mediating effect of consumer debt and debt problems among emerging adults. *Harm Reduction Journal, 15*, 1–11.

Orford, J. (2001). *Excessive Appetites: A Psychological View of Addictions*. New York, NY: John Wiley & Sons.

Petrović, Z. K., Peraica, T., Kozarić-Kovačić, D., & Palavra, I. R. (2022). Internet use and internet-based addictive behaviours during coronavirus pandemic. *Current Opinion in Psychiatry, 35*(5), 324–331.

Puiras, E., Cummings, S., Oliver, C., & Mazmanian, D. (2022). Add to cart: Problematic shopping and behavioral addictions in a Canadian sample during the COVID-19 pandemic. *Canadian Journal of Addiction, 13*(3), 26–35.

Purba, A. K., Thomson, R. M., Henery, P. M., Pearce, A., Henderson, M., & Katikireddi, S. V. (2023). Social media use and health risk behaviours in young people: Systematic review and meta-analysis. *BMJ, 383*, e073552.

Renu, N. (2021). Technological advancement in the era of COVID-19. *SAGE Open Medicine, 9*, 20503121211000912.

Rose, S., & Dhandayudham, A. (2014). Towards an understanding of Internet-based problem shopping behaviour: The concept of online shopping addiction and its proposed predictors. *Journal of Behavioral Addictions*, *3*(2), 83–89.

Savolainen, I., Oksanen, A., Kaakinen, M., Sirola, A., Zych, I., & Paek, H. J. (2021). The role of online group norms and social identity in youth problem gambling. *Computers in Human Behavior*, *122*, 106828.

Savolainen, I., Sirola, A., Vuorinen, I., Mantere, E., & Oksanen, A. (2022a). Online communities and gambling behaviors: A systematic review. *Current Addiction Reports*, *9*(4), 400–409.

Savolainen, I., Vuorinen, I., Sirola, A., & Oksanen, A. (2022b). Gambling and gaming during COVID-19: The role of mental health and social motives in gambling and gaming problems. *Comprehensive Psychiatry*, *117*, 152331.

Shanahan, L., Steinhoff, A., Bechtiger, L., Murray, A. L., Nivette, A., Hepp, U., ... & Eisner, M. (2022). Emotional distress in young adults during the COVID-19 pandemic: Evidence of risk and resilience from a longitudinal cohort study. *Psychological Medicine*, *52*(5), 824–833.

Sirola, A., Kaakinen, M., Savolainen, I., & Oksanen, A. (2019). Loneliness and online gambling-community participation of young social media users. *Computers in Human Behavior*, *95*, 136–145.

Sirola, A., Kaakinen, M., Savolainen, I., Paek, H. J., Zych, I., & Oksanen, A. (2021a). Online identities and social influence in social media gambling exposure: A four-country study on young people. *Telematics and Informatics*, *60*, 101582.

Sirola, A., Savela, N., Savolainen, I., Kaakinen, M., & Oksanen, A. (2021b). The role of virtual communities in gambling and gaming behaviors: A systematic review. *Journal of Gambling Studies*, *37*(1), 165–187.

Sirola, A., Nuckols, J., Nyrhinen, J., & Wilska, T. A. (2022). The use of the Dark Web as a COVID-19 information source: A three-country study. *Technology in Society*, *70*, 102012.

Sirola, A., Nyrhinen, J., & Wilska, T. A. (2023a). Psychosocial perspective on problem gambling: The role of social relationships, resilience, and COVID-19 worry. *Journal of Gambling Studies*, 1–19.

Sirola, A., Nyrhinen, J., Nuckols, J., & Wilska, T. A. (2023b). Loot box purchasing and indebtedness: The role of psychosocial factors and problem gambling. *Addictive Behaviors Reports*, *18*, 100516.

Spicer, S. G., Fullwood, C., Close, J., Nicklin, L. L., Lloyd, J., & Lloyd, H. (2022). Loot boxes and problem gambling: Investigating the 'gateway hypothesis'. *Addictive Behaviors*, *131*, 107327.

Topor, L., & Shuker, P. (2020). Coronavirus conspiracies and dis/misinformation on the Dark Web. *E-International Relations*, *9*, 2020.

Tuominen, J., Rantala, E., Reinikainen, H., & Wilska, T. A. (2023). Modern-day socialization agents: The connection between social media influencers, materialism, and purchase intentions of Finnish young people. *The Journal of Social Media in Society*, *12*(1), 21–48.

Tyrväinen, O., & Karjaluoto, H. (2022). Online grocery shopping before and during the COVID-19 pandemic: A meta-analytical review. *Telematics and Informatics*, *71*, 101839.

Vosoughi, S., Roy, D., & Aral, S. (2018). The spread of true and false news online. *Science*, *359*(6380), 1146–1151.

Vuorre, M., Zendle, D., Petrovskaya, E., Ballou, N., & Przybylski, A. K. (2021). A large-scale study of changes to the quantity, quality, and distribution of video game play during a global health pandemic. *Technology, Mind, and Behavior*, 2(4). https://doi.org/10.1037/tmb0000048.

Wang, Q., Kou, Z., Du, Y., Wang, K., & Xu, Y. (2022). Academic procrastination and negative emotions among adolescents during the COVID-19 pandemic: The mediating and buffering effects of online-shopping addiction. *Frontiers in Psychology, 12*, 789505.

Wang, Y., McKee, M., Torbica, A., & Stuckler, D. (2019). Systematic literature review on the spread of health-related misinformation on social media. *Social Science & Medicine, 240*, 112552.

Weise, K. (2021, April 29). Amazon's profit soars 220 percent as pandemic drives shopping online. *The New York Times*. https://www.nytimes.com/2021/04/29/technology/amazons-profits-triple.html.

Wilska, T. A., Sirola, A., Nuckols, J., & Nyrhinen, J. (2021). A year of COVID-19 in three countries: A study on the effects of the COVID-19 pandemic on everyday life, consumption and digital behaviour in Finland, Sweden, and Great Britain. *YFI julkaisuja*, (9).

Worthy, S. L., Jonkman, J., & Blinn-Pike, L. (2010). Sensation-seeking, risk-taking, and problematic financial behaviors of college students. *Journal of Family and Economic Issues, 31*, 161–170.

Yee, A. Z., & Sng, J. R. (2022). Animal crossing and COVID-19: A qualitative study examining how video games satisfy basic psychological needs during the pandemic. *Frontiers in Psychology, 13*, 800683.

Zamboni, L., Carli, S., Belleri, M., Giordano, R., Saretta, G., & Lugoboni, F. (2021). COVID-19 lockdown: Impact on online gambling, online shopping, web navigation and online pornography. *Journal of Public Health Research, 10*(1), jphr-2021.

Zhao, H., Tian, W., & Xin, T. (2017). The development and validation of the online shopping addiction scale. *Frontiers in Psychology, 8*, 735.

14. AI magnified inequalities: bias, (un)fairness and discrimination resulting from the use of AI-based technologies in the education sector

Maris Männiste and Andra Siibak

14.1 INTRODUCTION

The adoption of artificial intelligence (AI) or automated decision-making-based systems within the education sector experienced a significant increase during the COVID-19 pandemic, which created momentum for the educational technology (Edtech) industry. At the time, many AI-based systems, such as biometrics- and behavioural analytics-based AI technologies, as well as algorithms for standardizing results, started to be used by educational institutions around the world due to the need to find solutions for emergency remote teaching (Moore & Hodges, 2023). Yet, many of these technologies have by now emerged as a "new normal" (Selwyn et al., 2023, p. 152) in the present-day education sector and are heralded as "the potential solution to – nearly – every 'problem'" in the classroom (Chounta et al., 2022, p. 726). In fact, within the education sector, AI-based systems are often viewed as transformative and emancipatory technologies, which are able to provide insightful new knowledge and thereby operate as quick fixes to many different problems, from learning motivation and student engagement, to students' attendance and mental health. In short, it is expected that AI-based systems are "able to remedy obsolete and dysfunctional education systems that are limiting people's potential" (Davies et al., 2021, p. 551).

However, beneath the narratives of techno-solutionism (Morozov, 2013), which is eagerly promoted by Edtech, growing concerns about ethics and privacy (Lameras & Arnab, 2022), as well as the in-built (un)fairness and bias of AI-based technologies, have been voiced by many. In particular, scholarship has convincingly revealed that AI systems are biased across the intersection of gender (cf. Hall & Ellis, 2023 for systematic literature review) and race (cf. Intahchomphoo & Gundersen, 2020 for systematic review), leading to the exacerbation of existing inequalities in society. All the above concerns are also evident in the context of datafied education (Baker & Hawn, 2021), where the decisions to adopt AI systems are frequently driven by organizational, political, or financial goals. Questioning the decisions leading to the adoption of AI-based systems within education is thus crucial not only because public educational institutions are becoming increasingly "dependent upon private educational platforms and data-extractive infrastructures for many key functions, processes and operations" (Williamson et al., 2023, p. 561); but also owing to the growing number of AI-incidents (OECD, 2024), that is, events that have created actual harm, resulting from the use of AI-based systems.

In this chapter, we will zoom in on two specific AI incidents from the field of education that occurred during the COVID-19 pandemic: the algorithmic exam-grade moderation case from the United Kingdom, and the uptake of online proctoring software by higher education institutions around the world to combat academic dishonesty. The selected cases illustrate both the reliance on private providers' technologies at the time of emergency remote learning and the limited understanding that policymakers and governmental bodies have about AI systems. However, students who are both knowingly and unknowingly affected by these decisions and processes have limited opportunities to challenge or respond to them.

Considering that news media play an important role in shaping socio-technical imaginaries – that is, "collectively held and performed visions of desirable futures" (Jasanoff, 2015, p. 9), shared on a specific topic, we decided to make use of the critical discursive psychology approach (Locke & Budds, 2020) to explore how the AI-based systems adopted in the case of these incidents were publicly scrutinized by a variety of stakeholders – politicians, educators, students, and tech vendors – and reported by the international press. Our final sample comprises 320 news articles covering the adoption of online proctoring technologies and 329 news stories about the algorithmic exam-grade moderation case, published from 2020–2021.

In this chapter, we will consolidate the findings from our empirical news media analysis with the aim of showcasing how the adoption of AI-based systems was experienced by students. Furthermore, we will argue that the decisions to issue exam grades with an algorithm and to combat academic dishonesty with proctoring technologies caused harm to the most vulnerable and marginalized of the students, widening social inequalities.

14.2 BIAS IN AI AND RELATED HARMS

Previous scholarship has indicated that the use of new AI-based systems and the resulting collection of data can lead to many risks for already marginalized groups. In fact, research (Benjamin, 2019; Noble, 2018) stipulates that AI-based technologies often amplify the already existing vulnerabilities (McQuillan, 2022) leading to algorithmic bias, "understood broadly as the tendency of computational systems to reflect the inequalities and injustices that exist in society" (Perrotta, 2023, p. 187). Extensive media coverage surrounding the many discriminatory episodes (i.e., Correctional Offender Management Profiling for Alternative Sanctions (COMPAS) in the US; System Risk Indication (SyRI) in the Netherlands; teenage-pregnancy prediction algorithm in Argentina) that have occurred in different cultural contexts and life spheres has broken the illusion of neutral and fair algorithms and has highlighted the effects of automation on marginalized groups. As Airoldi (2021, p. 43) puts it: "the myth of algorithmic neutrality clashes with the reality where algorithmic decisions or results are experienced as racist, sexist, or classist".

But why are algorithms or AI systems biased? The bias in the systems, on the one hand, "originates from the *deus ex machina*, that is, the encoded choices and intentions of those involved in the development of the systems" (Airoldi, 2021, p. 43). All machine creators see the world through the lens of their own background and experience. Airoldi (2021, p. 43) emphasizes that, as a result, developers may also ignore or underestimate the implications of pursuing specific computational goals or selecting certain parameters and variables.

Still, the design features are just one source of bias and the easiest to correct, for which the creators of the technology are accountable (Airoldi, 2021). Design features account for only a minor part of the possible errors and misbehaviours of the platform algorithms and AI agents (Campolo et al., 2017). When the results of algorithmic systems are racist, classist, or sexist, they often fall under what is called "data bias" (Olteanu et al., 2019) or "training data bias" (Danks & London, 2017). Such biases embedded in the technology cannot be considered as "mere glitches", but rather should be treated as structural biases that cannot be fixed easily "with a quick code update" (Broussard, 2023, p. 4). These biases are built upon "historical patterns of discrimination and classification which often construct harmful representations of people based on perceived differences" (Whittaker et al., 2018, p. 25) often resulting in allocative harms preventing access to basic resources such as education, jobs, medical help, or loans. Hence, as stated by Hoffmann (2020), allocative harms can lead to the erosion of citizenship and human rights. In fact, Benjamin (2019) uses the notion of "the New Jim Code" to refer to the ways AI-based technologies reproduce and exacerbate historical inequalities while being proposed as neutral and objective.

Discussions of algorithmic bias in education have also been complicated due to the overlapping meanings of the term (Blodgett et al., 2020). For instance, it is applied to showcase various instances of unfairness of automated systems, which, as Baker and Hawn (2021) argue, fit statistical or technical definitions of bias. Therefore, the questions of bias also cannot be separated from the questions of what constitutes fairness (Kizilcec & Lee, 2022).

Fairness in the context of AI can be described as "the absence of any prejudice or favoritism toward an individual or a group based on their inherent or acquired characteristics. An unfair algorithm is one whose decisions are skewed towards a particular group of people" (Mehrabi et al., 2019, p. 1). Thus, fairness in the context of AI systems "concerns what and whose values are embedded into the technology, who is excluded as a result, and whether this is justified in the eyes of the law, in light of the moral principles and in the eyes of the individuals who may be affected by the way in which given technology works" (Porayska-Pomsta et al., 2023, p. 578). This makes fairness an essential measure through which unwanted consequences are countered (Hutchinson & Mitchell, 2019) when AI systems are developed. But as Hartong and Breiter (2020, p. 55) indicate, fairness is also a "general political rationale to foster equality in education systems, ranging from resource allocation to assessment programs, rankings, and accountability systems". This means that there are different types of fairness (e.g., individual vs. group fairness, outcome vs. process fairness) at play, resulting in conflicting interpretations of what is and what is not fair (Porayska-Pomsta et al., 2023, p. 579). The different understandings of fairness make it crucial for AI systems developers, as well as policymakers who opt to use these technologies, to outline the ideas that influence how fairness is incorporated into AI systems. In fact, for years, scholars have highlighted the need to "open the black box" (Pasquale, 2015), that is, to make the inner workings of the AI systems auditable and transparent. But transparency should, as Poryska-Pomsta et al. (2023, p. 579) emphasize, also include making the system's operations explainable to the users. In short, it is not enough when the inner workings of the system are made more transparent and trustworthy, it is also necessary to make the designers and decision-makers "accountable for the outcomes resulting from their systems' use" (Porayska-Pomsta et al., 2023, p. 579). The problem is that when pedagogical expertise is shifted away from practitioners in proprietary systems (Thompson et al., 2023, p. 248), as in the case of AI systems developed by private companies, the inner workings of the systems become trade secrets and expertise is displaced from classrooms and schools to the sites of software development.

Below, we showcase two AI incidents where the use of AI-based systems in the education sector led to unjust discrimination against vulnerable student groups.

14.3 CASE STUDIES

14.3.1 The (Un)fair Algorithm: the Use of Algorithms in Standardizing Exam Grades in the UK

In education, the problem of harmful bias entered mainstream discourse in 2020 when, due to the COVID-19 pandemic, students in the UK were unable to take their A-Level exams (taken by 16–18-year-olds), which are necessary for being admitted to university. The Office of Qualifications and Examinations Regulation (Ofqual) rejected the option to award students' grades based on teacher assessment and proposed a new and seemingly more objective and accurate solution – the "grade standardization algorithm" (cf. Smith, 2020) – contrasting the idea with the supposedly subjective judgement of teachers (the centre-assessed grade, CAG). Such a decision vividly illustrates the typical narratives about the impartiality and objectivity associated with datafication and the use of AI-based technologies as well as portraying "the perceived mistrust of policymakers of teachers' capacity for professional judgment" (Kelly, 2021, p. 14). Ofqual's aim was to prevent the possible inflation of teacher-provided grades, through the use of an algorithm that would standardize the results based on past patterns of achievement. According to an article in *The Guardian*, different stakeholders, from scholars working on AI to parents (Pidd, 2020), voiced their concerns about this solution; however, these concerns were not taken seriously. Furthermore, as the components of the exam-grade algorithm were not announced before the grades were revealed, algorithmic discrimination remained black-boxed and hidden from public view.

Only after the grades were announced and it became apparent that the algorithm had downgraded nearly 36 per cent of the grades lower than their teachers' original A-Level assessment would have been (Kelly, 2021), did the public learn that the algorithm had displayed bias by systematically downgrading the performance of state-funded schools (Gulson et al., 2021) contributing to the reproduction of educational inequity (Allegretti, 2020). Hence, in this particular case, the historical data became important in deciding the future of current students – grades downgrading was impacted more by the schools' previous performance and less by the students' own performance during a particular year. This meant that students with high-achieving results did not get the results they deserved because of the school's previous performance. For some, especially for students from the larger state-run schools that have "historically educated a larger proportion of Black, Asian and Minority Ethnic (BAME) students" (Wachter et al., 2021, p. 737) it meant that the algorithm standardized grades were in some cases almost 40 per cent lower than the teacher's predicted grades. Although a record proportion of school-leavers also received a top A* grade and there was a considerable increase in the proportion of As awarded, improvement in A* and A grades was most significant for students attending independent schools (up by 4.7 per cent), while the grades of students from state comprehensive schools and sixth-form colleges showed a very small improvement (0.3 per cent) (Ofqual, 2020, p. 136). The application of the algorithm can therefore be seen as yet another case of "digital redlining" that is, "a set of education policies, investment decisions, and IT practices that actively create and maintain class boundaries through strictures that discriminate against specific groups" (Gilliard & Culik, 2016). Redlining as a concept originated from the historical practice in housing and real estate in the US, which meant that primarily black residents would be denied resources such as loans in their neighbourhoods (Bessette, 2023). As Gilliard and Culik (2016) argue, redlining created differences in physical access to schools, libraries, and home ownership. Digital redlining integrated into Edtech showcases the same kind of discriminatory results, as students quoted in the news articles referred to the situation as a "postcode lottery" (see, for example, Mohdin, 2020) indicating that their exam results did not really "correlate with their own efforts" (see, for example, Ferguson & Savage, 2020). Reports by the news media (see, for example, Adam, 2020), described how being graded unfairly by the algorithm resulted in university rejections for students coming from more marginalized communities and lower socio-economic backgrounds. Although Ofqual officials did, in later interviews (see, for example, Hazell, 2020), confirm that a downward adjustment was more likely to happen for students with lower socio-economic backgrounds, they attributed this trend to teachers being more generous towards them (Kelly, 2021). As Kelly (2021) notes – this argument had no evidential basis.

Only after the students' protests and public outrage at the unfair results and the too-difficult appeals process, which would not help to resolve students' worries related to university places and possible legal action, was the UK government was forced to take what news media reported as an "embarrassing", or "humiliating", U-turn (see, for example, Weale et al., 2020). This meant that about 15,000 students who had been rejected by their first-choice university due to the lower algorithm-generated grades were able to reapply using their teacher's estimate (Kelly, 2021). Hence, as different commentators of the case noted in the media, the problem was not just the use of the algorithm in standardizing grades or the unfair results, but also the process that followed. Although an appeals process was assigned, teachers interviewed by the press voiced their concerns, claiming that disadvantaged students may not even think about appealing, that is, challenging the decision made by the system (see, for example, Ferguson, 2020).

Wong (2020) argues that it is essential to acknowledge the relationship between different stakeholders' interests and fairness measures and see algorithmic fairness measures as inevitably favouring specific interests. In the A levels case, government and Ofqual officials initially defended the use of algorithms and often claimed in the media that the chosen "model is the fairest possible way to calculate grades for larger cohorts" (Lee, 2020). However, as noted by Shaw and Nisbet (2021, p. 9) "calculated grades which seemed fair in aggregate later seemed unfair to teachers and school leaders when they saw the implications for their own schools and their own students".

The underlying idea behind using the algorithm, therefore, reflects certain policy decisions, that is, controlling the number of university applicants through grade moderation, thus also making fairness a political rationale that depends on the specific aims and needs of those governing and less on those governed through the AI-based systems (Hartong & Breiter, 2020).

14.3.2 AI-automated Discrimination: the Use of Online Proctoring Software by Higher Education Institutions

At the time of the pandemic, there was another case that resulted in intense scrutiny for privacy shortcomings and bias towards marginalized groups: the adoption of online proctoring software.

Online proctoring software started to be used by higher education institutions to validate students' identities and monitor students' activities at the time of online exams (Kharbat & Abu Daabes, 2021). Even though the first of these systems started to be used two decades ago, reliance upon online proctoring software sharply increased during the first two years of the pandemic when non-remote options were simply not available (see Khalil et al., 2022 for a systematic literature review). For example, the findings of a poll among EDUCAUSE community institutions (*N*=312; 294 from the US) carried out in April 2020 suggest that, at the time of the study, 54 per cent of the surveyed educational institutions were using such technologies and another 23 per cent were considering using them (Grajek, 2020) regardless of the fact that proctoring technologies were accused of intrusions of privacy and poor adoption of social and ethical values (e.g., transparency, justice, fairness, trust, responsibility, accountability, etc.) (cf. Nigam et al., 2021).

Online proctoring software makes use of biometrics to link observable gestures and facial expressions to suspicious behaviour and suspected cheating in the hopes of protecting academic integrity. In fact, media reported that "the software turns students' computers into powerful invigilators" (Feathers & Rose, 2020) – the microphone is recording all the noise from the room, the webcam is monitoring eye and head movements, and algorithms log how often the student presses the keys or moves their mouse. Any behaviour the algorithm deems suspicious is flagged in the hopes of catching cheaters and thereby helping to ensure the "validity" of university degrees (Selwyn et al., 2023). At the same time, a comprehensive overview of available online proctoring software (Arnó et al., 2021) and empirical studies from 2015–2021 (Nigam et al., 2021) refer to a set of technical problems these technologies have in accurately detecting and reporting student malpractices. Problems with false positives or baseless "flagging" of inappropriate behaviour during exams result in extensive manual controlling (cf. Amigud et al., 2018).

In many universities across the globe, students signed petitions and protested against the adoption of such technologies (cf. Selwyn et al., 2023). Media reports highlighted that online proctoring technologies were not only claimed to be "creepy and unacceptable", but the students signing the petitions also considered the use of such technologies to be a "blatant violation" of privacy (Kelly, 2020). Although such a "sense of discomfort" was "broadly echoed by students" (Khalil et al., 2022, p. 1600), news media showcased how such criticisms were brushed aside by many higher education institutions (Kimmons & Veletsianos, 2021).

At the same time, students' concerns about equity, security, privacy, accessibility, and bias in online proctoring technologies were not voiced without evidence. Our analysis of international news media articles revealed many problems students, especially those coming from more marginalized communities, had to face due to the uptake of online proctoring technologies. For instance, news media articles revealed that students from lower socio-economic backgrounds were often simply unable to fulfil the necessary technological requirements needed for taking the exam online as they did not have the financial means to buy a decent laptop or a webcam; nor had they the means to ensure that they had a stable and fast internet connection that was necessary for the software to work properly (Honorlock is not…, 2020). Media reported, that on some occasions, students were notified two days prior to their bar exam that they were not allowed to take exams on their desktop computers (Patrice, 2020) leading to frustrated students claiming that such technologies are "really measuring a person's generational wealth and not their knowledge of the law" (Feathers & Rose, 2020).

Our analysis of news media articles indicates that the use of online proctoring software created discriminatory consequences across multiple identities. In particular, news media reported many instances where students with black or brown skin, or had East Asian, South Asian, or Central American origin (Brown, 2020) experienced problems with the technology, as the software was calibrated to detect white skin but struggled to recognize non-white faces. Journalists described that on those occasions, students had to come up with uncomfortable solutions, for example, keeping the light shining directly into their faces for the whole duration of the exam (Feathers & Rose, 2020), or had to turn their heads towards the window for the software to recognize them (Metz, 2020). Such problems reveal the in-built bias of facial recognition systems due to the fact that images of under-represented groups have been less available in the training sets of tech vendors (Yoder-Himes et al., 2022).

News media described that proctoring technologies were also discriminatory against disabled students, including neurodivergent students and students with ADHD symptoms (Kelly, 2020). Reports by privacy activists also revealed that the use of online proctoring technologies triggers anxiety in students who have chronic anxiety, post-traumatic stress, and attention deficit disorders, as well as in students who have motor tics, cerebral palsy, or who are autistic (Brown, 2020). In addition, the keystroke-analysis feature of the software was noted to flag inconsistent typing as a possible sign of cheating rather than consider it to be caused by some medical conditions or physical disabilities (Swauger, 2020b). Such kinds of instances however, can lead to "data violence" (Hoffmann, 2020), that is, disproportionate harm, against communities that are already marginalized or non-dominant (Swauger, 2020a, n.p) as the systems unfairly penalize students for things that are out of their control.

14.4 DISCUSSION

In this chapter, we have illustrated through two case studies – exam-grade standardization through an algorithm and the uptake of online proctoring software – various ways in which the adoption of different AI-based systems that were propagated as objective, neutral, and fair actually leads to magnifying inequalities within the education sector.

Findings from our international news media analysis related to the algorithmic grade case not only highlighted the issues around different types of fairness, but also revealed how the lack of proper transparency and accountability measures ended up creating an AI incident, harming students coming from more marginalized groups. Furthermore, in addition to the students who had to deal with the consequences of digital redlining, teacher autonomy and expertise were seriously minimized. In fact, the case serves as a grim example of how naive and at the same time technochauvinistic the socio-technical imaginaries held by the ruling powers were about the neutrality and objectivity of AI-based systems. In practice, such imaginaries paved the way for pedagogical decisions and duties, such as grading, which have traditionally been seen as one of the key responsibilities of a teacher, to be assigned to a machine.

News media analysis related to the use of online proctoring software in higher education institutions across the world provided many examples of how these AI-based systems ended up causing (data) harm across multiple identities. Through the use of online proctoring technologies, students' computers were turned into surveillance machines that tended to reduce the user's humanity, not to mention their agency and autonomy, to a mere collection of data points. In fact, news articles revealed how the students ended up being represented by opaque algorithms assigning them into measurable types, whose bodies could be "encoded as either normal or suspicious" and whose behaviours could be considered as either "safe or threatening" (Swauger, 2020a). Furthermore, even though students were actively voicing their concerns and protesting against the use of online proctoring technologies that were not only violating their privacy but were also racist, ableist, and discriminatory, they were unable to opt out from such surveillance and were often powerless to create the necessary change.

In both of the analysed cases, students from more vulnerable groups experienced the most data violence from the use of AI-based systems (Olteanu et al., 2019). Considering that algorithmic classifications "regulate and govern an individual's position in society by shaping the directions and futures that become available to them, thus potentially producing new forms of digital and social inequality" (Mascheroni & Siibak, 2021, p. 176), such AI incidents can be life-altering. Being rejected by a university due to low grades generated by an algorithm, or failing an exam after being incorrectly flagged for suspicious behaviour, are experiences that may start to shape the opportunities, well-being, and the futures available to the young.

The two cases chosen for this chapter were actively covered and discussed in international news media, and in many cases it was this public scrutiny that pushed the universities and the UK government to overturn their previously made decisions. However, many AI-based technologies may never reach the point when they are publicly questioned by data activists or the general public. Hence, there is a need for new legislative mechanisms that would help to ensure that the AI systems being developed and used in educational contexts are shaped to really meet the needs of the students and teachers, rather than the technology sector interested in earning a profit.

14.4.1 Practical Implications

The number of different AI incidents, that is, AI system implicated in alleged harm or near-harm events, is expected to grow in the coming years. Thus, there is a crucial need to actively invest in automated auditing and public oversight bodies that would help to develop and deploy safe, responsible, trustworthy, and unbiased AI systems. Furthermore, as argued by Meredith Broussard (2023, p. 172) "if we are building AI systems that intervene in people's lives, we need to maintain and inspect and replace the systems in the same way we maintain and inspect and replace bridges and roads". Hence, rather than investing billions into inventing new automated systems, it is about time that societies started funding initiatives that aim to improve the already existing technologies, as they not only replicate age-old stereotypes and reproduce inequalities, but also create new socio-technical vulnerabilities for future generations.

14.4.2 Avenues for Future Research

Analysis of news media articles revealed that students' voices were rarely used when covering the cases. Rather, issues around the uptake of AI-based technologies in education were mainly commented on by adults, for example, school officials, policymakers, tech vendors, teachers, or parents. Hence, it is not only crucial to enable students' voices be heard in public discussions, but it is also necessary to engage in research that would enable young people to express their opinions, socio-technical imaginaries, and experiences related to the use of AI-based technologies in the education sector. Furthermore, future research should also aim to explore the experiences of young people who have lived through AI incidents like the ones discussed in this chapter. Such empirical research is much needed to fill the void of youth-led, scientifically informed, and fair information about young people's everyday encounters with AI-based systems.

REFERENCES

Adam, K. (2020, August 18). The U.K. used an algorithm to estimate exam results. The calculations favored elites. *Washington Post*. https://www.washingtonpost.com/world/europe/the-uk-used-an-algorithm-to-estimate-exam-results-the-calculations-favored-elites/2020/08/17/2b116d48-e091-11ea-82d8-5e55d47e90ca_story.html (accessed 4 March 2024).

Airoldi, M. (2021). *Machine Habitus: Toward a Sociology of Algorithms*. John Wiley & Sons.

Allegretti, A. (2020, August 13). A-level results: Government accused of "baking in" inequality with "boost" for private schools. *Sky News*. https://news.sky.com/story/35-of-a-level-results-downgraded-by-one-grade-figures-reveal-12048251 (accessed 22 May 2024).

Amigud, A., Arnedo-Moreno, J., Daradoumis, T., & Guerrero-Roldan, A.-E. (2018). An integrative review of security and integrity strategies in an academic environment: Current understanding and emerging perspectives. *Computers & Security*, *76*, 50–70. https://doi.org/10.1016/j.cose.2018.02.021.

Arnò, S., Galassi, A., Tommasi, M., Saggino, A., & Vittorini, P. (2021). State-of-the-art of commercial proctoring systems and their use in academic online exams. *International Journal of Distance Education Technologies (IJDET)*, *19*(2), 55–76.

Baker, R. S., & Hawn, A. (2021). Algorithmic bias in education. *International Journal of Artificial Intelligence in Education*, 1–41.

Benjamin, R. (2019). *Race After Technology: Abolitionist Tools for the New Jim Code*. Polity Press.

Bessette, L. S. (2023). Digital redlining, minimal computing, and equity. In S. Köseoğlu, G. Veletsianos, & C. Rowell (Eds), *Critical Digital Pedagogy in Higher Education* (pp. 111–130). AU Press.

Blodgett, S. L., Barocas, S., III, H. D., & Wallach, H. (2020). Language (technology) is power: A critical survey of "bias" in NLP. In *Proceedings of the 58th Annual*

Meeting of the Association for Computational Linguistics (pp. 5454–5476). https://doi.org/10.18653/v1/2020.acl-main.485.

Broussard, M. (2023). *More than a Glitch. Confronting Race, Gender, and Ability Bias in Tech*. MIT Press.

Brown, L. X. Z. (2020, November 16). *How Automated Test Proctoring Software Discriminates Against Disabled Students*. Centre for Democracy & Technology. https://cdt.org/insights/how-automated-test-proctoring-software-discriminates-against-disabled-students/ (accessed 22 May 2024).

Campolo, A., Sanfilippo, M. R., Whittaker, M., & Crawford, K. (2017). *AI Now 2017 Report*. AI Now Institute at New York University.

Chounta, I. A., Bardone, E., Raudsep, A., & Pedaste, M. (2022). Exploring teachers' perceptions of artificial intelligence as a tool to support their practice in Estonian K-12 education. *International journal of artificial intelligence in education*, *32*(3), 725–755.

Danks, D., & London, A. J. (2017, August). Algorithmic bias in autonomous systems. *Ijcai*, *17*, 4691–4697.

Davies, H. C., Eynon, R., & Salveson, C. (2021). The mobilisation of AI in education: A Bourdieusean field analysis. Sociology, 55(3), 539–560. https://doi.org/10.1177/0038038520967888.

Feathers, T., & Rose, J. (2020, September 24). Students are rebelling against eye-tracking exam surveillance tools. *VICE*. https://www.vice.com/en/article/n7wxvd/students-are-rebelling-against-eye-tracking-exam-surveillance-tools (accessed 25 May 2024).

Ferguson, D. (2020, August 11). UK schools fear backlash from parents over "unfair" A-level exam results. *The Guardian*. https://www.theguardian.com/education/2020/aug/08/poorer-students-will-suffer-in-unfair-a-level-grading-model-uk-parents-warn (accessed 25 May 2024).

Ferguson, D., & Savage, M. (2020, August 16). Autumn term chaos feared over exam resits and appeals. *The Guardian*. https://www.theguardian.com/education/2020/aug/16/autumn-term-chaos-feared-over-exam-resits-and-appeals (accessed 4 March 2024).

Gilliard, C., & Culik, H. (2016, May 11). Filtering content is often done with good intent, but filtering can also create equity and privacy issues. *Common Sense Education*. www.commonsense.org/education/articles/digital-redlining-access-and-privacy (accessed 25 May 2024).

Grajek, S. (2020). Educause COVID-19 QuickPoll results: Grading and proctoring. *EDUCAUSE Review*. https://er.educause.edu/blogs/2020/4/educause-covid-19-quickpoll-results-grading-and-proctoring#fn3 (accessed 25 May 2024).

Gulson, K., Benn, C., Kitto, K., Knight, S., & Swift, T. (2021, November 21). Algorithms can decide your marks, your work prospects and your financial security. How do you know they're fair? *The Conversation*. https://theconversation.com/algorithms-can-decide-your-marks-your-work-prospects-and-your-financial-security-how-do-you-know-theyre-fair-171590 (accessed 25 May 2024).

Hall, P., & Ellis, D. (2023). A systematic literature review of socio-technical gender bias in AI algorithms. *Online Information Review*, *47*(7), 1264–1279. https://doi.org/10.1108/OIR-08-2021-0452.

Hartong, S., & Breiter, A. (2020). Between fairness optimization and "inequalities of dataveillance": The emergence and transformation of social indices in German school monitoring and management. In S. Grek, C. Maroy, & A. Verger (Eds.), *World yearbook of education 2021: Accountability and datafication in the governance of education* (pp. 54–71). Routledge.

Hazell, W. (2020, August 13). A-level results 2020: 39% of teacher predicted grades downgraded by algorithm amid calls for U-turn. *iNews*. https://inews.co.uk/news/education/a-level-results-2020-grades-downgraded-algorithm-triple-lock-u-turn-result-day-578194 (accessed 25 May 2024).

Hoffmann, A. L. (2020). Terms of inclusion: Data, discourse, violence. *New Media & Society, 23*(12), 3539–3556.https://doi.org/10.1177/1461444820958725.

Honorlock is not the key to proctoring (2020, September 28). *The Mercury*. https://utdmercury.com/honorlock-is-not-the-key-to-proctoring/ (accessed 25 May 2024).

Hutchinson, B., & Mitchell, M. (2019, January). 50 years of test (un)fairness: Lessons for machine learning. In FAT '19: *Proceedings of the Conference on Fairness, Accountability, and Transparency* (pp. 49–58), ACM. https://doi.org/10.1145/3287560.3287600.

Intahchomphoo, C., & Gundersen, O. E. (2020). Artificial intelligence and race: A systematic review. *Legal Information Management, 20*(2), 74–84. https://doi.org/10.1017/S1472669620000183.

Jasanoff, S. (2015). Future imperfect: Science, technology, and the imaginations of modernity. In S. Jasanoff & S.-H. Kim (Eds.), *Dreamscapes of modernity: Sociotechnical imaginaries and the fabrication of power* (pp. 1–33). University of Chicago Press.

Kelly, A. (2021). A tale of two algorithms: The appeal and repeal of calculated grades systems in England and Ireland in 2020. *British Educational Research Journal, 47*(3), 1–15.

Kelly, J. (2020, September 25). Students are pushing back against proctoring surveillance apps. *Electronic Frontier Foundation*. https://www.eff.org/deeplinks/2020/09/students-are-pushing-back-against-proctoring-surveillance-apps (accessed 25 May 2024).

Khalil, M., Prinsloo, P., & Slade, S. (2022). In the nexus of integrity and surveillance: Proctoring (re)considered. *Journal of Computer Assisted Learning, 38*, 1589–1602. https://doi.org/10.1111/jcal.12713.

Kharbat, F. F., & Abu Daabes, A. S. (2021). E-proctored exams during the COVID-19 pandemic: A close understanding. *Education and Information Technologies, 26*, 6589–6605. https://doi.org/10.1007/s10639-021-10458-7.

Kimmons, R., & Veletsianos, G. (2021, February 23). Proctoring software in higher ed: Prevalence and patterns. *Educause Review*. https://er.educause.edu/articles/2021/2/proctoring-software-in-higher-ed-prevalence-and-patterns (accessed 25 May 2024).

Kizilcec, R. F., & Lee, H. (2022). Algorithmic fairness in education. In W. Holmes & K. Porayska-Pomsta (Eds), *Ethics in Artificial Intelligence in Education* (pp. 174–202). Taylor & Francis.

Lameras, P., & Arnab, S. (2022). Power to the Teachers: An Exploratory Review on Artificial Intelligence in Education. *Information, 13*(1), 14. https://doi.org/10.3390/info13010014.

Lee, G. (2020, August 12). Did England exam system favour private schools? *Channel 4 News*. https://www.channel4.com/news/factcheck/factcheck-did-england-exam-system-favour-private-schools (accessed 25 May 2024).

Locke, A., & Budds, K. (2020). Applying critical discursive psychology to health psychology research: A practical guide. *Health Psychology and Behavioral Medicine*, 8, 234–247.

Mascheroni, G., Siibak, A. (2021). *Datafied Childhoods. Data Practices and Imaginaries in Children's Lives*. New York, United States of America: Peter Lang Verlag.
McQuillan, D. (2022). *Resisting AI: An anti-fascist approach to artificial intelligence*. Policy Press.
Mehrabi, N., Morstatter, F., Saxena, N., Lerman, K., & Galstyan, A. (2019). A survey on Bias and fairness in machine learning. *ArXiv E-Prints*, arXiv:1908.09635. https://arxiv.org/abs/1908.09635
Metz, R. (2020, August 29). Online school means online tests, along with computerized surveillance. *CNN Business*. https://edition.cnn.com/2020/08/29/tech/online-school-test-surveillance/index.html (accessed 25 May 2024).
Mohdin, A. (2020, August 14). Downgraded A-level students urged to join possible legal action. *The Guardian*. https://www.theguardian.com/education/2020/aug/13/downgraded-a-level-students-urged-to-join-possible-legal-action (accessed 25 May 2024).
Morozov, E. (2013). To save everything, click here: The folly of technological solutionism. Public Affairs.
Moore, S. L., & Hodges, C. B. (2023). Emergency remote teaching. In *EdTechnica: The Open Encyclopedia of Educational Technology*. EdTech Books. https://edtechbooks.org/encyclopedia/emergency_remote_teaching.
Nigam, A., Pasricha, R., Singh, T., & Churi, P. (2021). A systematic review on AI-based proctoring systems: Past, present and future. *Education and Information Technologies*, *26*(5), 6421–6445.
Noble, S. U. (2018). *Algorithms of Oppression*. New York University Press.
OECD (2024). Defining AI incidents and related terms. *OECD Artificial Intelligence Papers*, 16. OECD Publishing. https://doi.org/10.1787/d1a8d965-en (accessed 25 May 2024).
Ofqual (2020). Awarding GCSE, AS and A levels, in summer 2020: Interim report. www.gov.uk/government/publications/awarding-gcse-as-a-levels-in-summer-2020-interim-report (accessed 25 May 2024).
Olteanu, A., Castillo, C., Diaz, F., & Kıcıman, E. (2019). Social data: Biases, methodological pitfalls, and ethical boundaries. *Frontiers in Big Data*, *2*, 13.
Pasquale, F. (2015). *The Black Box Society: The Secret Algorithms that Control Money and Information*. Harvard University Press.
Patrice, J. (2020, September 17). Two days before the deadline, bar exam declares desktop computers forbidden. *Above The Law*. https://abovethelaw.com/2020/09/two-days-before-the-deadline-bar-exam-declares-desktop-computers-forbidden/ (accessed 25 May 2024).
Perrotta, C. (2023). Advancing data justice in education: Some suggestions towards a deontological framework. *Learning, Media and Technology*, *48*(2), 187–199.
Pidd, H. (2020, August 15). "Punishment by statistics": The father who foresaw A-level algorithm flaws. *The Guardian*. https://www.theguardian.com/education/2020/aug/14/punishment-by-statistics-the-father-who-foresaw-a-level-algorithm-flaws (accessed 25 May 2024).
Porayska-Pomsta, K., Holmes, W., & Nemorin, S. (2023). The ethics of AI in education. In *Handbook of Artificial Intelligence in Education* (pp. 571–604). Edward Elgar Publishing.
Selwyn, N., O'Neill, C., Smith, G., Andrejevic, M., & Gu, X. (2023). A necessary evil? The rise of online exam proctoring in Australian universities. *Media International Australia*, *186*(1), 149–164.

Shaw, S. D., & Nisbet, I. (2021). Attitudes to fair assessment in the light of COVID-19. *Research Matters: A Cambridge Assessment publication*, *31*, 6–21.

Smith, H. (2020). Algorithmic bias: Should students pay the price? *AI & Society*, *35*, 1077–1078. https://doi.org/10.1007/s00146-020-01054-3.

Swauger, S. (2020a). Our bodies encoded: Algorithmic test proctoring in higher education. In J. Stommel, C. Friend, & S. M. Morris (Eds), *Critical Digital Pedagogy. A Collection*. PressBooks. https://pressbooks.pub/cdpcollection/chapter/our-bodies-encoded-algorithmic-test-proctoring-in-higher-education/.

Swauger, S. (2020b, November 7). Remote testing monitoried by AI is failing the students forced to undergo it. *NBC News*. https://www.nbcnews.com/think/opinion/remote-testing-monitored-ai-failing-students-forced-undergo-it-ncna1246769 (accessed 25 May 2024).

Thompson, G., Gulson, K. N., Swist, T., & Witzenberger, K. (2023). Responding to sociotechnical controversies in education: A modest proposal toward technical democracy. *Learning, Media and Technology*, *48*(2), 240–252.

Wachter, S., Mittelstadt, B., & Russell, C. (2021). Bias preservation in machine learning: The legality of fairness metrics under EU non-discrimination law. *West Virginia Law Review*, *123*(3), 735–790. https://dx.doi.org/10.2139/ssrn.3792772.

Weale, S., Stewart, H., & Elgot, J. (2020, August 26). Ofqual head Sally Collier resigns over exams fiasco. *The Guardian*. https://www.theguardian.com/education/2020/aug/25/ofqual-head-sally-collier-resigns-over-exams-fiasco (accessed 25 May 2024).

Whittaker, M., Crawford, K., Dobbe, R., Fried, G., Kaziunas, E., Mathur, V., West, S. M., Richardson, R., Schultz, J., & Schwartz, O. (2018). *AI Now Report*. https://ainowinstitute.org/AI_Now_2018_Report.pdf (accessed 25 May 2024).

Williamson, B., Eynon, R., Knox, J., & Davies, H. (2023). Critical perspectives on AI in education: Political economy, discrimination, commercialization, governance and ethics. In B. du Boulay, A. Mitrovic, & K. Yacef (Eds.), *Handbook of Artificial Intelligence in Education* (pp. 553–570). Edward Elgar Publishing.

Wong, P. H. (2020). Democratizing algorithmic fairness. *Philosophy & Technology*, *33*, 225–244.

Yoder-Himes, D. R., Asif, A., Kinney, K., Brandt, T. J., Cecil, R. E., Himes, P. R., Cashon, C., Hopp, R. M. P., & Ross, E. (2022). Racial, skin tone, and sex disparities in automated proctoring software. *Frontiers in Education*, *7*, 881449. doi:10.3389/feduc.2022.881449.

15. Making a life through digital (in)securities: the entanglement of risks and skills in teen refugees' digital lives

Myria Georgiou, Leen d'Haenens, Alia Zaki, Verónica Donoso and Emilie Bossens

15.1 INTRODUCTION

By choice or circumstances, for many young refugees, life is digital life. Yet, for several, this condition comes with many contradictions and asymmetries, not least as digital communications converge opportunities and risks to manage the insecurities associated with being a refugee and with being young. This chapter explores what it means to lead a digital life as a refugee *and* as a teenager in Europe and to navigate a life full of insecurities and uncertainties with the support or hinderance of online tools, resources, and connections. Three research questions drive the discussion. How do young refugees mobilize digital tools, resources, and connections to manage everyday, ontological, and digital insecurities? Are they equipped with digital skills to effectively manage fast and sometimes traumatic changes in their lives? Or do they develop these skills out of their need to learn, communicate, and comprehend their transnational condition's uncertainties? These questions are raised at a time when the conditions of so many young refugees' digital lives are full of challenges – from managing family separation online to navigating risky encounters on social media.

Based on the analysis of our research findings, we aim to understand the two core dimensions of these challenges: heightened insecurities and digital risks. Specifically, the lives of many young refugees, we will show, are torn by insecurities, not least ontological insecurity – what Giddens (1991) describes as individuals' sense of lacking safety, belonging, and stability when they experience a world full of risks and uncertainties. At the same time, as many of them lead digital lives, they constantly have to learn how to balance between the opportunities of online connectivity and the many risks that this same connectivity presents, namely disconnection, surveillance, and marginalization (Awad & Tossell, 2021; Dhoest, 2020).

The chapter navigates this complex landscape of young refugees' digital lives. It aims to contribute to understanding the ambivalence of digitization in contexts of vulnerability, by paying particular attention to the ways agency is mobilized to manage precarity, while also recognizing the limits and obstacles to young people's attempts to manage precarity. Drawing on 96 interviews and five creative workshops with young refugees of different backgrounds in Athens, Brussels, and London, we discuss the entanglement of vulnerability and agentive capabilities when young refugees develop their skills in the context of their digital lives.

15.2 LITERATURE REVIEW: DIGITAL SKILLS FOR MANAGING TASKS, OR LIFE?

There is a growing body of research that investigates the role of digitization in the experience of young refugees, and with migrants more generally. This research has been systematic but largely divided along two main strands. The first focuses on digital skills' role in securing or hindering young people's integration, primarily in educational and work environments in the countries of settlement. The second strand is more interested in digital communications' role in contexts of acute refugee precarity and intense ontological insecurities, such as in camps and spaces of transit. We discuss them in turn, recognizing their contributions but also limits.

15.2.1 Functional Skills for Inclusion in Education and Work

Primarily focusing on the specific benefits of digitization for post-migration integration, a significant body of research examines refugee lives in destination countries. More specifically, this line of research has examined how digital pathways can compensate for interruption in educational and professional development, enhancing integration in schools and work environments (cf. Beadle, 2014; Koehler & Schneider, 2019). This research has often responded to the repeatedly recorded inequalities young refugees and migrants more generally face in terms of educational and professional achievement. As a result of "inadequate language support, failure to acknowledge and meet psychosocial and emotional needs, racism, discrimination and cultural insensitivity", according to a report by the Centre for Multicultural Youth (Beadle, 2014, p. 15), young refugees and migrants suffer from high levels of school dropout and continuous high rates of unemployment. Many face huge obstacles, as employment is tied to demand for advanced digital skills, while their own digital skills are often limited to communication skills, such as participating in social network sites (Eynon & Geniets, 2015).

Obstacles, research shows, are both systemic and cultural. Assumptions made within educational institutions about linguistic, digital, and broader levels of literacy alienate many, making it more difficult to catch up with required skills at school and work (Eynon & Geniets, 2015). Martinovic et al. (2019), among others, discuss the importance of digital skills on young people's advancement in education and work, but also in sustaining self-esteem and overall social well-being, while Kneer et al. (2019) remind us of the ambivalent outcomes of peer-to-peer support. Their study, in Dutch schools, revealed how many young refugees benefited from learning with their classmates but suffered from low self-esteem when their peers assumed they had certain digital skills, and so commented negatively or mocked them when they realized that they did not.

This literature also shows how this constituency, even at the heart of Europe, still suffers from restrictions to one of the most fundamental tools needed to participate in education, work, and society more generally: internet access. Given the levels of deprivation many refugees face, something we also observed in Athens and London, it is not uncommon for some to depend on public WiFi for connectivity. Yet, public WiFi comes with many restrictions, for example, slow and timed connections that limit opportunities for the cultivation of particular skills and the navigation of complex computer functionalities.

The contradictory force of digital communications that furthers everyday insecurities is another concern raised in the literature. Being online offers links to opportunities for educational engagement and employment; however, as with other teenagers, digital connectivity can amplify the risks to young people's well-being, including through cyberbullying, technology addiction, and isolation (Hawi & Samaha, 2016). Other findings also show that unlimited access, "where internet use is both facilitated and rewarded" (Eynon & Geniets, 2015, p. 473) often affects digital skills in contradictory ways, with excessive use often negatively impacting negatively school performance. These concerns have also been addressed with proposals for enhancing peer support to compensate for educational, emotional, and residential insecurity and change (Kneer et al., 2019), and for developing bespoke and professional digital training, such as in the case of the coding schools for refugees in Germany (Rushworth & Hackl, 2021).

Overall, this first strand of scholarship privileges functional digital skills as crucial to securing integration in education and work. However, digital communications do not exist in a vacuum but in specific socio-spatial contexts, where many struggle with emotional and psychological insecurities, referred to below as ontological insecurities (Silverstone, 1993).

15.2.2 Critical Digital Skills for Managing Extreme Precarity

The contradictory force of digital communications is at the heart of the second strand of scholarship, often described as *digital migration research*. This research engages with refugees' digital lives under conditions of heightened insecurities and intense change. A threefold overarching argument drives digital migration research: digital skills often develop in the context of, and as a response to, conditions of acute insecurity for life and well-being; refugees mobilize their agentive capabilities to understand and manage consequent adversity, often through their digital practices; and under those conditions digital risks and opportunities inevitably converge (cf. Alencar, 2020; Gillespie et al., 2018; Wall et al., 2016). This line of research builds on interdisciplinary understandings of the situatedness of communication in societies riven by risks and in individual lives shaped through insecurities.

Silverstone (1993), drawing on Giddens' influential conceptualization of ontological security, has highlighted the dual role of the media in exacerbating ontological insecurities; media, he argues, constantly remind audiences of the risks that surround them, yet, at the same time, the media contain these insecurities by turning risks into a staple of everyday life, especially as they constantly appear on our screens. Livingstone (2013) takes the discussion further by noting that risks are not only represented digitally but are increasingly also manifested digitally (e.g., through cyberbullying). Yet, she continues, risks do not pre-empt negative effects; the agentive engagement of young people with platforms and networks can produce a range of different, and always contextual, outcomes (Livingstone, 2013).

Drawing on these critical scholarly interpretations of communication, agency, and change, digital migration research has delved into examining the different outcomes of digital connectivity in conditions of vulnerability. This research can be organized along two subthemes that privilege critical digital skills and their restrictions: the first subtheme examines internet access and communication in conditions of precarity, and the second focuses on migration's own role in individuals' development of critical digital skills and literacies.

The ordinariness of ontological and everyday insecurities and the consequent development of digital skills to manage them is extensively discussed under the first subtheme. Alencar (2020) note that young refugees learn how to navigate smartphones and social media to sustain relationships of intimacy and care across the multispatial and unstable context of their transnational lives, while Georgiou and Leurs (2022) explain how smartphones become individual pocket-size archives linking the dots of their lives between the past and present and here and there. Communication skills are often learnt on the move and mobilized on smartphones to support a sense of ontological security when everything is uncertain, with many refugees noting that their phones are "most essential" and "my best friend" (Alencar, 2020). Scholarship on digital migration has also shown that the smartphone often becomes the par excellence technology to manage ontological insecurities in contradictory ways: creating opportunities for sustaining relations that are fundamental to well-being, for example, intrafamilial connections, while at the same time enhancing risks, for example, exposure to surveillance by senior family members and even by powerful actors such as the state (Chouliaraki & Georgiou, 2022; Dhoest, 2020).

Taking into account the many obstacles that young refugees face, digital skills often develop through subversion. In camps, Wall et al. (2016) show, many young refugees develop hacking skills to gain internet access despite limited or fully restricted formal access. Through hacking, many engage in what Beadle (2014) refers to as "ICT Wayfaring" – a tactic for gathering information and offering peer-to-peer assistance and sharing hacked passwords. Additionally, the most digitally literate members of the community mediate others' internet access in the camp, a practice that indicates how digital skills become collective and communally shared – with certain individuals, usually young people, acting as brokers of knowledge and communication (Aziz, 2022).

For Wall et al. (2016, p. 240), these conditions reflect refugees' information precarity, which they define as "the condition of instability that refugees experience in accessing news and personal information, potentially leaving them vulnerable to misinformation, stereotyping, and rumours that can affect their economic and social capital". Conditions of information precarity often limit refugees' abilities to develop critical digital skills and literacies. This is even more so the case for women and poorer refugees, those detached from networks of support, but also for those overexposed to false or dangerous information, as well as to surveillance, as is often the case with young refugees, especially men.

Critical digital skills and digital literacy, the second subtheme in digital migration research, emphasize that refugees need access to a range of crucial information, often information that makes a difference between life or death, between well-being and extreme suffering (Gillespie et al., 2018). As they rarely have access to reliable and official sources of information to support safe journeys and resettlement, many depend on information only available through informal and sometimes dubious digital networks. Given the stakes of such information – for example, getting access to irregular routes of escape, especially when legal routes are unavailable, contacting smugglers while in transit (Imani Giglou et al., 2022) – refugees have to *learn by doing* (De Coninck et al., 2023), evaluating as they go reliable and unreliable information, information that sometimes makes a difference between life and death (Gillespie et al., 2018).

Skills to navigate and verify information extend beyond those moments of crucial decision-making. Not many populations feel that regional and global news affairs directly affect their lives, yet this is often the case with refugees. Many refugees regularly compare and contrast media originating in different regional and linguistic zones (Chouliaraki & Georgiou, 2022) and, in the process, they lose trust in mainstream media, as social media become their primary sources of information (Wall et al., 2016).

15.2.3 Challenging the Functional Versus Critical Skills Divide

The above approaches respectively focus on applied digital skills for educational and work achievements after resettlement, and on digital skills for survival and management of conditions of extreme insecurity, especially in camps and while in transit. While this research is informative and important, it has two significant limitations. First, in both lines of investigation, we see a fragmentation of migrant digital experience and a division between skills for specific purposes (e.g., education) and skills for life (e.g., survival). Yet, our research has shown that, in conditions of vulnerability, these different skills are intertwined and develop simultaneously as young refugees turn to digital communications to compensate for limited access to stable structures (e.g., a home and schooling), as these are connected to practical needs but also emotional, financial, and ontological security. Effectively, this scholarly fragmentation represents the second limitation of research that either ignores ontological security (the first strand) or examines it only in acute conditions of crisis (the second strand).

As will be shown below, young refugees mobilize their agency to flexibly and contextually develop skills necessary for managing a range of insecurities that coexist – everyday, ontological, and digital. Thus, and against the dominant binary of functional versus critical skills through which existing literature is organized, we propose a *triangulated approach* to digital skills that does not start from specific problems (e.g., online learning, family connectivity), but from the observed embeddedness of digital communications in young refugees' lives. This is a bottom-up approach that is guided by participants' own reflections. As we have repeatedly heard, participants constantly move from functional to critical skills, doing different things online, all at the same time. What drives their engagement with digital communication? Effectively, their attempts to manage different insecurities frame their lives as young people and as uprooted migrants. Listening to their own realities, we organize our analysis by focusing on three dimensions of insecurity: everyday insecurities, ontological insecurities, and digital insecurities. We discuss our findings after a brief discussion of our methodology.

15.3 DATA AND METHODOLOGY

This chapter draws on research conducted with young refugees between 2021 and 2022 in urban settings (where advanced digital infrastructures and concentrated migration converge) across three European urban settings: Athens, Brussels, and London. Specifically, we conducted 96 interviews and five creative workshops with refugees, each engaging between eight and 12 participants. We primarily engaged with Afghan and Syrian youth, and secondarily with others who have experienced violent uprooting in Sudan, Venezuela, Yemen, and elsewhere. All participants were between 14 and 19 years old at the time. The chapter draws primarily on interview thematic analysis and secondarily on observational notes collected during the creative workshops. The interviews provided space for individual reflections, while the creative workshops enabled the nonlinear and collective production of deliberations in playful and visual ways (Sharp, 2024).

The participant groups shared certain commonalities but also differences across countries; thus, no direct comparisons between locations are claimed. For example, they share age and experiences of uprooting and resettlement, and regular, although uneven, access to digital communications. Yet, the particular conditions of their lives vary significantly. Athens' participants come mostly from Afghanistan and the Middle East and are conditioned by the EU's bordering policies of limited and ephemeral recognition (Chouliaraki & Georgiou, 2022); London participants are the most diverse group in terms of ethnicity and experience; most have access to some basic resources, such as education, but many suffer from acute isolation and high levels of poverty. Participants in Brussels have all experienced family separation, as they are unaccompanied minors, and face extreme levels of change, as they all come from rural Afghanistan. Similarly, the policy frameworks for integration and refugee recognition, or their lack, are uneven between the three countries. Belgium offers the most consistent reception infrastructure for young refugees, Greece lacks any adequate framework for integration, and the UK provides some structures of support but only meagre resources for living a decent life. This diversity might not allow for neat comparisons between countries and groups but, most importantly, it generates a qualitative and rich narrative that maps out the complexities that drive and frame many young refugees' digital lives. It also provides an important reminder that there is no one kind *of refugee*, in the same way that there is no one kind *of teen*, or a single European policy of migration, for that matter.

15.4 FINDINGS: BEING YOUNG AND ONLINE – LIVING WITH AND THROUGH DIGITAL INSECURITIES

As we will see below, many participants depend on digital communications to manage everyday and ontological insecurities as refugees and as young people who want to learn, have fun, and ground themselves in the world. Whether they manage to achieve these aims or not is a complicated and contextual matter. Here, we organize the findings in a systematic way that aims to capture how (i) engagement with digital communications is always agentive; (ii) it relates to participants' identities as people who are both refugees and young; and (iii) effectively and practically, young participants develop digital skills in order to manage insecurities that range from the ordinary (e.g., sustainable everyday communication), to the ontological (e.g., sense of belonging), and to the digital (e.g., online learning of life skills). We discuss these findings under three subheadings: Digital skills for managing everyday insecurities; Digital skills within/for ontological insecurities; Well-being with and through digital (in-)securities.

15.4.1 Digital Skills for Managing Everyday Insecurities

For most participants, a life of uprooting and resettlement has come with experiences of trauma, fast and sometimes violent change, and significant levels of uncertainty about the future. These widely shared vulnerabilities, however, take their meanings within the geographies of transnational, digital lives; specifically, participants experience vulnerabilities not only as refugees but also as teenagers, that is, as people at a transitional life stage that comes with its own challenges. Thus, living with insecurity becomes ordinary, though not untroubled, as seen below.

> The good thing [about social media] is that you can communicate with people, and you can get closer to them more, you can speak to them wherever you are. The bad thing is that on social media apps like Instagram and TikTok, you see how nice people's lives are and how nice their bodies are and you feel upset and wonder why I don't have the same life or look like this. This has really affected me. (Syrian, female, 16, London)

We see here how, for this young woman, the entanglement of digital risks and opportunities is lived experience. On the one hand, she recognizes how social media enables connection, but, on the other, how, as in the case of many young people, it also exacerbates mental pressures (Livingstone et al., 2022).

In everyday life, it also becomes ordinary for many participants to seek security through online connectivity. Digital skills, we repeatedly saw, are shaped precisely through the need to connect and learn – both for education and for pleasure, both about what is happening in their countries of origin but also their country of settlement. YouTube, in particular, becomes both informative and upsetting, precisely in the way Silverstone (1993) notes – that is, with media exacerbating but also containing anxieties in a world torn by risks.

> I try learning general information and knowing about new things, so I follow channels that give new information and tell stories or narrate religious events. I also follow channels that talk about natural disasters that are happening in the world. (Syrian, female, 17, London)

Digital communications are often the most immediately available resources that many turn to in order to navigate new environments. For example, many participants use their smartphones to resist marginalization and isolation, developing in the process skills, both for digital and for urban navigations (Gillespie et al., 2018).

> We just tried to explore the area we live in, Peckham. We tried searching for shops, we don't know anything here. We found a gym so we joined it, we found a library that we also joined, we tried to make a membership but they told us we needed an ID; we still haven't gotten one yet. (Syrian, male, 14, London)

In conditions of adversity, we see how young refugees turn to digital resources to build autonomy (Alencar, 2020). Many participants across the three cities, for example, use YouTube tutorials to compensate for gaps in their schooling and to learn different languages. In these different ways, platforms become self-educating resources, compensating for lack of formal access to vital resources, for example, sustainable and uninterrupted schooling.

> Here in Greece the way I use the phone has changed. I also use it to learn and read a lot of lessons and look for different ones that interest me. I want to study political science, learn about law and philosophy and I'm looking for how I can do this after school. (Afghan, male, 16, Athens)

These different examples of *learning by doing* (De Coninck et al., 2023) confirm the entanglement of functional and critical skills necessary for progressing in school but also for making life plans. It also shows how finding a place in the world means addressing different elements of one's identity:

> I watch YouTube for video tutorials to study Dutch and stay abreast of Pashto music. (Afghan, male, 15, Brussels)

And for some, digital communications are also about developing specific skills.

> Well, the computer was always interesting for me. I remember when I was 13, I saw a book in a school library lab that was a C sharp programming language. I started doing that and I've managed to finish the book in one day and then I started to do C sharp programming. (Iranian, male, 15, London)

Digital skills, we also found, are not only developed by doing, but also by *doing together*; they often reveal a collective agency of learning from each other. Learning together is both a common and vital practice for many participants who find themselves within the unending temporariness of the migrant experience; within contexts of interruption and uncertainty, peer networks often substitute for missing structured and stable systems of support. This does not mean that collective knowledge is produced in an untroubled manner. It is often subjected to gatekeeping and interveillance within the group (Christensen & Janson, 2015), and it reveals the intersectionality of transnational lives. Networks of co-learning and sharing information and technologies, such as smartphones, develop in the context of restrictions and opportunities relating to gender, ethnic, and religious affiliations, as well as local and transnational cultural norms. For example, we noticed that rural Afghan boys in Belgium depended on older male figures for their digital knowledge.

15.4.2 Digital Skills within/for Ontological Insecurities

"Ordinary life" for most participants is a life of violent interruptions as a result of war and borders, and thus, inevitably, of exacerbated insecurities compared with many other young people. Within this context, digital connectivity compensates for the interruption of intimate relationships crucial to young people's well-being – for example, with parents, siblings, friends – and for the continuity of their sense of being (Giddens, 1991; Silverstone, 1993) – for example, when unable to make plans for the future. Yet, it is not only being a refugee but also being young that frames the intertwinement of digital connectivity with ontological security. As we see in this section, facing a range of risks – from isolation to bullying – is at the same time experienced and managed digitally.

The words of a Sudanese young man seeking to sustain a sense of ontological security in conditions of extreme adversity capture the centrality of digital communications in doing so:

> Firstly, in Malta, I was not on social media and I did not have any contact with my family as I couldn't access them. I made a Facebook [account] to try to contact my family, but I was sadly unsuccessful because they don't use social media and don't even have internet. I am keeping it in case I can find them one day. (Sudanese, male, 18, London)

These words painfully reveal adversity but also hope, borders but also agency, when a young man aims to sustain a sense of continuity in a life interrupted (Giddens, 1991).

This is no exception, as we often heard many participants explaining how they learnt to negotiate connections between *here and there* and between the continuity of past and present.

> I call my parents in Afghanistan once a month, more contact is not possible because of frequent network problems. I have more contact with my friends in Belgium, France and Afghanistan, once or twice a week. On my way to Belgium, I did not have a smartphone with me. Sometimes I asked if I could use a friend's phone to reach my family via Facebook or Messenger. Each time, I made sure I logged off. (Afghan, male, 15, Brussels)

Learning to use social media strategically is not merely practical but crucial to feeling safe in one's self and within one's fast-changing world. This becomes an intense process when many participants have to learn how to economize their limited resources (time, connectivity) and, inevitably, to maximize digital opportunities while, at the same time, learning how to protect themselves from digital risks (Livingstone, 2013). As we will see in the next section, participants consequently develop an, often fragile, sense of security, not outside or against risks but *within and through* them.

15.4.3 Well-Being with and through Digital (In-)securities

For many young refugees, there is no choice but to depend on digital communications to fulfil needs that cannot be met otherwise, either because of social exclusion or because of the restrictive governance of migration that separates them from loved ones and destabilizes conditions of life. Inevitably then, and even more so than for other young people, digital communications become fundamental to managing everyday needs, ontological needs, even existential needs. Thus, everyday and ontological insecurities, as we see in this subsection, are often channelled through digital practices. In and through them, we observed, young refugees learn to live with insecurities, though not without being exposed to many risks.

The words of a young Syrian woman capture most vividly the entanglement of risks and opportunities. She explained how she had to throw away her smartphone – as smartphones' GPS technology, which can be life-saving for many refugees, is now increasingly used by states to push refugees back and away from Europe.

> Because we came in an illegal way, and they told us that we shouldn't have a phone. In general, not just when we crossed the border. I left it back in Lebanon

> and nobody had a phone. At the border, they have something to monitor, a radar, and they detect mobile phones. Nobody had a mobile phone. (Syrian, female, 18, Athens)

Skills for managing risks are always contextual, Livingstone (2013) reminds us. In the case above, as with others, this means that evaluating the pros and cons of owning technologies under conditions of extreme precarity can turn into a matter of life or death.

Alongside the complex conditions of migration, many also expressed awareness of the risks associated with their digital lives as young people, for example, unsolicited contact on social media. Most said that they block people they do not know; some, especially girls, explained that they had experiences of strangers approaching them online and have discussed the matter with their friends. Let us remind ourselves that a lot of what young refugees do online is similar to what other young people do. Perhaps it is as young people or as refugees, or most likely because their identities emerge at the juncture of developmental and life transition, that most seem to understand these risks as manageable, though still anxiety-generating.

> Yeah, like imagine on Insta I met some Arab boy and he said he lived in London. So, he asked me to meet, just to be friends because he doesn't have Arab friends. So, I said, okay. I found out he's a good guy, but sometimes it's not good to have friends from social media; you don't know if they're good or bad. (Syrian, male, 16, London)

The younger participants were more likely to rely on intuition (Livingstone et al., 2022) to deal with digital risks, for example, observing communication markers in social media such as names, and being cautious with accepting friend requests from strangers, or simply ignoring them and scrolling on. Digital risks were also evaluated in the wider context of individuals' risk-ridden lives. Some were even surprised that we wanted to know so much about risks online, as they considered them minuscule compared with other risks they are exposed to. But for others, digital risks are still unsettling and a cause of insecurity in their own right.

> I saw on social media, like two weeks ago, that there's a lot of teenagers that are committing suicide because of social media, that's like the downside of it. (Senegalese, male, 18, London)

The way some participants deal with the experience of losing a phone, an experience that many refugees encounter many times in their lives and often in risky conditions (e.g., when travelling along irregular and dangerous routes, cf. Alencar, 2020; Gillespie et al., 2018) is in itself revealing. This is an experience that uncovers the deep ontological dependence of many upon their phone for sustaining significant relations, but also a sense of continuity in their identity.

> So, listen to what has happened to my phone. And I lost it, and they took it from me, and my screen is broken because it fell from me. And every time I was sad, I felt that everyone would leave … My mom did not want to get me a new one to learn to be careful. I also stopped eating and told her, "if you do not get me a new one, I will not eat again". After a couple of days she got me one. (Congolese, female, 17, Athens)

Losing a phone is part of a life that is full of losses; thus, this material loss can trigger significant distress. Against loss and trauma, often, it is the ordinary use of smartphones that restores calm.

> I use Facebook just to see some photos of my friends. Some are in Germany, some are in France. If I have free time, I play football and I post some photos of myself on Facebook too. (Afghan, male, 14, Brussels)

This section showed how seeking security is regularly channelled through digital practice. In fact, quite often, we discovered, digital practice becomes a pathway to sustaining continuity in young people's identities, not away from risks, but often through them.

15.5 DISCUSSION: SKILLS FOR LIVING WITH AND THROUGH DIGITAL (IN-)SECURITIES

Young refugees' digital lives are full of risks, we saw, but also opportunities to manage insecurities associated with everyday needs to socialize and learn, all the way to ontological needs to find security, stability, and intimacy in lives interrupted and restarted. What our analysis reveals is that, for many young refugees, digital life is shaped, on the one hand, through the ambivalence of the reassurances and opportunities embedded in digital practice and, on the other, the risks and gaps that are associated with doing so much of what they do online – socializing, connecting with family, learning, and developing a sense of self. We organized this chapter's discussion through three research questions, which we return to here to highlight our key findings before concluding.

The first question asked how young refugees mobilize digital tools, resources, and connections to manage everyday, ontological, and digital insecurities. Our findings showed that we need to address this question at the juncture of age and transnational identities. We saw again and again that most participants depend on smartphones to lead social lives, like so many other teenagers. This dependence is communicational and brings them many joys. At the same time, it generates anxieties, not least for fear of missing out and for not meeting the standards of beauty and wealth promoted across platforms. We are thus reminded that this group is not essentially different from their peers and that their complex and intersectional identities emerge within wider social and technological contexts. At the same time, participants' digital practices need to be understood in dialogue with forced migration, which carries two specific conditions: first, the unsettling realities of uprooting, migration policies, and interrupted intimacies that sometimes turn the smartphone into a life-changing and life-threatening technology; and, second, the social marginalization that deprives many of having regular access to digital knowledge and the tools necessary for developing complex skills associated with education and work.

The chapter's second question aimed to examine whether young refugees are equipped with the digital skills to effectively manage fast and sometimes traumatic changes in their lives. The research findings discussed above illustrate how most participants *learn by doing*, and do so not only as individuals but also collectively, developing, as a consequence, skills that are often shared and sometimes surveilled within hierarchical relations of gender, age, and migration governance. This represents one of our important findings as it illustrates, on the one hand, how individual and collective agency contest limited and unreliable access to resources for achieving educational and social stability, and, on the other, how agency and young refugees' resilience is impressive but can rarely fully compensate for persistent conditions of marginality and exclusion.

The third question aimed to examine what young refugees developed digital skills for. What we found is that, unlike the case of many young people, for whom digital skills represent integral elements of formal learning as much as experience, for young refugees, digital skills emerge mostly out of experiential necessity. Many participants have gone through trauma, uprooting, family separation, and uncertainty for the future – all while also living in conditions of marginality, often in poverty or in shelters for unaccompanied minors. For them, digital skills reflect much more than digital literacy; they represent a necessity required to navigate a very fast-changing world full of risks and vulnerabilities. While this does not mean that digital communications equip them with all the tools and skills to navigate the difficult conditions they so often face, they have become absolutely essential for their navigation through life.

This chapter contributes both to research on digital skills in the contexts of vulnerability and to digital migration scholarship. First, we saw that a life mediated through digital insecurities is generative of digital skills. For young refugees, learning to do so much of what they do online is a necessity if they are to avoid isolation and manage marginality. This also means that their digital skills are at the same time functional and critical, and they take different meanings in varied contexts. For example, communication and interaction skills (Helsper et al., 2021) might be basic and functional skills that enable sociality online, but they can become foundational skills for well-being, as interrupted lives need the groundedness of communication for sustaining ontological security.

15.5.1 Practical Implications

Our research with young refugees, a group often experiencing vulnerability and marginalization, reminds us that digital skills need to be understood as skills for life. Understanding digital skills as opportunities for enhancing educational and employment achievement is not enough. For those at the margins, living transnational lives torn by insecurity, digital skills are as much about finding pathways to education and work as they are about generating agency and finding ways to ground oneself in a world of multiple exclusions, misrepresentations, and risks to well-being, sometimes even for life itself.

15.5.2 Avenues for Future Research

This chapter emphasizes the need to examine digital skills not only in relation to the intended purposes of the skills but also through the social and cultural meanings that are recreated and reconfigured in digital environments. Second, the chapter points to the scholarly need to examine migration as not only challenging for refugees but also as generative of agency and resistance. Individually and collectively, the participants of this study mobilize different skills against marginality, isolation, and trauma, as we have seen, through their engagement with digital communications. These agentive acts, we note, could be recognized as skills for life. The triangulated analysis presented in this chapter hopes to contribute to understanding youth skills intersectionally, collectively, contextually, and as they are situated within individuals' and social groups' experiences of insecurity, and marginality, but also experiences of hope.

REFERENCES

Alencar, A. (2020). Mobile communication and refugees: An analytical review of academic literature. *Sociology Compass*, *14*, e12802. https://doi.org/10.1111/soc4.12802.

Awad, I., & Tossell, J. (2021). Is the smartphone always a smart choice? Against the utilitarian view of the "connected migrant". *Information, Communication & Society*, *24*(4), 611–626. https://doi.org/10.1080/1369118X.2019.1668456.

Aziz, A. (2022). Rohingya diaspora online: Mapping the spaces of visibility, resistance and transnational identity on social media. *New Media & Society*, *26*(9), 5219–5239. https://doi.org/10.1177/14614448221132241.

Beadle, S. (2014). *Facilitating the Transition to Employment for Refugee Young People*. Centre for Multicultural Youth.
Chouliaraki, L., & Georgiou, M. (2022). *The Digital Border: Migration, Technology, Power*. NYU Press.
Christensen, M., & Janson, A. (2015). Complicit surveillance, interveillance, and the question of cosmopolitanism. Towards a phenomenological understanding of mediatization. *New Media & Society*, *17*(9), 1473–1491.
De Coninck, D., Vissenberg, J., Joris, W., & d'Haenens, L. (2023). Perceived discrimination and digital inequalities among children and young people: Studying the multidimensional concepts of digital skills and digital knowledge. *Information, Communication & Society*.https://doi.org/0.1080/1369118X.2023.2205508.
Dhoest, A. (2020). Digital (dis)connectivity in fraught contexts: The case of gay refugees in Belgium. *European Journal of Cultural Studies*, *23*(5), 784–800. https://doi.org/10.1177/1367549419869348.
Eynon, R., & Geniets, A. (2015). The digital skills paradox: How do digitally excluded youth develop skills to use the internet? *Learning, Media and Technology*, *41*(3), 463–479.
Georgiou, M., & Leurs, K. (2022). Smartphones as personal digital archives? Recentring migrant authority as curating and storytelling subjects. *Journalism*, *23*(3), 668–689.
Giddens, A. (1991) *Modernity and Self-identity*. Polity.
Gillespie, M., Osseiran, S., & Cheesman, M. (2018). Syrian refugees and the digital passage to Europe: Smartphone infrastructures and affordances. *Social Media + Society*, *4*(1), 1–18.
Hawi, N. S., & Samaha, M. (2016). To excel or not to excel: Strong evidence on the adverse effect of smartphone addiction on academic performance. *Computers & Education*, *98*, 81–89. https:// doi.org/10.1016/j.compedu.2016.03.007.
Helsper, E., Schneider, L., Van Deursen, A., & Van Laar, E. (2021). *The Youth Digital Skills Indicator. Report on the Conceptualisation and Development of the ySKILLS Digital Skills Measure*. KU Leuven, Leuven: ySKILLS.
Imani Giglou, R., Buiter, A., Joris, W., Borowski, M., & d'Haenens, L. (2022). Understanding the role of social media on the road to social inclusion: The case of Syrian refugees in Belgium. In P. Tsatsou (Ed.), *Vulnerable People and Digital Inclusion: Theoretical and Applied Perspectives* (pp. 97–117). Palgrave Macmillan.
Kneer, J., Van Eldik, A., Jansz, J., Eischeid, S., & Usta, M. (2019). With a little help from my friends: Peer coaching for refugee adolescents and the role of social media. *Media and Communication*, *7*(2), 264–274.
Koehler, C., & Schneider, J. (2019). Young refugees in education: The particular challenges of school systems in Europe. *Comparative Migration Studies*, *7*(1), 28. https://doi.org/10.1186/s40878-019-0129-3.
Livingstone, S. (2013). Online risk, harm and vulnerability: Reflections on the evidence base for child Internet safety policy. *ZER: Journal of Communication Studies*, *18*(35), 13–28.
Livingstone, S., Stoilova, M., Stänicke, L. L., Jensen, R. S., Graham, R., Staksrud, E., & Jensen, T. K. (2022). *Young People Experiencing Internet-related Mental Health Difficulties: The Benefits and Risks of Digital Skills. An Empirical Study*. KU Leuven.
Martinovic, D., Freiman, V., Lekule, C. S., & Yang, Y. (2018). The roles of digital literacy in social life of youth. In *Encyclopedia of Information Science and*

Technology, Fourth Edition (pp. 2314–2325). IGI Global. https://doi.org/10.4018/978-1-5225-7659–4.ch009.

Rushworth, P., & Hackl, A. (2021). Writing code, decoding culture: Digital skills and the promise of a fast lane to decent work among refugees and migrants in Berlin. *Journal of Ethnic and Migration Studies, 48*(11), 2642–2658. https://doi.org/10.1080/1369183x.2021.1893159.

Sharp, R. (2024). *Refugee Voices.* Routledge.

Silverstone, R. (1993). Television, ontological security and the transitional object. *Media, Culture & Society, 15*(4), 573–598.

Wall, M., Otis Campbell, M., & Janbek, D. (2016). Syrian refugees and information precarity. *New Media & Society, 19*(2), 240–254.

16. Digital maturity: enabling a beneficial use of digital technologies by children and adolescents

Franziska Laaber, Teresa Koch, Arnd Florack and Marco Hubert

16.1 INTRODUCTION

Digital transformation is rapidly evolving and affecting numerous activities in our lives (Nyrhinen et al., 2022). The use of mobile information and communication technologies (ICTs), such as smartphones, tablets, and laptops, has become an integral part of daily activities, especially for children and adolescents (Livingstone et al., 2021). An increasing number of children regularly interact with mobile ICTs and participate in social networking sites such as Instagram, Snapchat, or TikTok, which provide access to a wide range of content and activities online (Bozzola et al., 2022; Kardefelt Winther et al., 2019). The acceptance and rapid adoption of novel ICT devices and functions (e.g., smart assistants, virtual and augmented reality, and chatbots based on generative artificial intelligence), and their seamless integration into children's daily lives, are constantly evolving (Laaber et al., 2023; Livingstone et al., 2018).

The increasing intrusion of technology into daily life has not come without scepticism. The pervasive integration of mobile technologies into young people's lives has prompted an extensive discourse on the potential risks and opportunities these technologies present (e.g., Bell et al., 2015). Concerns have been raised about the potential negative effects of heavy device use on young people's mental and physical well-being (Brautsch et al., 2023; Scott et al., 2017; Twenge & Campbell, 2018), their self-image (Kim et al., 2021; Vogel et al., 2015), and their online safety and privacy (Shin & Kang, 2016). Young people are particularly vulnerable to the temptations and content of digital technologies and social networks. For instance, popular social networking features such as "likes" or "shares" have been shown to activate the brain's reward system, encouraging repeated engagement (Meshi et al., 2015). At the same time, mobile technologies offer young people ways to connect with others (Valkenburg & Peter, 2011), gain knowledge (Gan et al., 2015), and express their identities (Huang et al., 2021).

With the rise of technological capabilities and new types of media, research has focused on the different skills and their underlying antecedents needed to use technology effectively while avoiding potential harm. The primary skills examined are media, information, and digital literacy (Helsper et al., 2020). These forms of literacy describe a person's ability to use and understand specific technologies (e.g., tablets) or media types (e.g., social media; Livingstone et al., 2021). The impact of different levels of media literacy and digital literacy among children and adolescents has been a particular focus of research (Buckingham, 2015). Children's and adolescents' ICT-related behaviour can be partly explained by factors such as usage patterns, sociodemographic variables, digital competence, parental mediation, and children's digital literacy levels (Helsper et al., 2020; Livingstone et al., 2018). However, explaining when ICT-related behaviours are beneficial vs. maladaptive for individuals requires more than just understanding media or digital literacy or ICT use patterns. Against this background, this research aims to explore the dimensions and implications of a new and more holistic approach to young people's use of digital technologies: digital maturity.

In this chapter, psychosocial maturity and self-determination theory (Bleidorn, 2015; Deci & Ryan, 2000; Greenberger et al., 1975) are used to conceptualize and define digital maturity as the "capabilities and attitudes that enable individuals to use digital technologies in ways that support their personal development (growth) and social integration (adjustment)" (Laaber et al., 2023, p. 2). Our definition of digital maturity is based on personality maturity but extends beyond it to include the skills and mindsets needed to successfully navigate the digital environment. Thus, this chapter illuminates the concept of children's digital maturity as the basis for a comprehensive taxonomy and applies it to establish a theoretical framework for explaining relationships between children's digital maturity and relevant determinants and outcomes.

16.2 CONCEPTUALIZATION OF DIGITAL MATURITY AND THEORETICAL BACKGROUND

Digital technology can have both positive and negative effects on children and young people. Potential disadvantages include excessive screen time (Paulich et al., 2021), digital addiction (Kim et al., 2021), privacy and safety concerns (Park, 2013), and risky social interactions such as cyberbullying (Bastiaensens et al., 2016). However, young people can also benefit from digital tools (Haddon & Livingstone, 2017). They use mobile devices and the internet for a variety of positive purposes, including self-expression and civic engagement, as well as schoolwork, gaming, video viewing, chatting, and content creation (Huang et al., 2021; Ólafsson et al., 2013). Moreover, daily interactions with the digital space promote the development of digital literacy (Livingstone & Helsper, 2010), access to education (Gan et al., 2015; Livingstone & Bober, 2004), and the creation of online social networks (Valkenburg & Peter, 2009). Given the potential risks and opportunities of digital technologies, there is a need for further research on the impact of digital technologies and social media on young people's well-being. In fact, a recent umbrella review (Valkenburg et al., 2022) revealed that the majority of studies found weak to inconsistent relationships between young people's use of social media and their well-being.

One reason for the inconsistent results might be that most studies in this area focus on either the positive or negative effects of digital technologies on young people. However, they do not provide a sufficient understanding of how young people can develop digital maturity – a kind of maturity that goes beyond simple technical competence and empowers them to use technology for their positive development and integration into social networks and society. Thus, our notion of digital maturity encompasses social integration in digital contexts as well as personal development.

16.2.1 Personality Maturity and Self-Determination as a Starting Point

In developing the concept of digital maturity, we were inspired by theories of personality development that emphasize the importance of growth and adjustment in people's psychosocial maturity (Staudinger & Kunzmann, 2005). Adjustment describes how well a person functions in society (Greenberger et al., 1975) and is linked to traits such as conscientiousness (e.g., organization, diligence, and reliability), agreeableness (e.g., compassion, politeness, and trust in others), and emotional stability (e.g., low emotional volatility and the ability to cope with difficult situations; Bleidorn, 2015). Conversely, growth refers to aspects of personality development such as self-improvement and self-determination that go beyond social integration; that is, a person's ability to make decisions and manage their life on their own (Deci & Ryan, 2000).

In our work, we view social adjustment and personal development as two interrelated facets of human development and apply these facets to the field of digital technology. To properly manage digital challenges and threats, digital maturity integrates elements of social adjustment and personal growth (Laaber, 2024). The ability of digital technologies to influence users, for example, through the design of social media applications, can be a threat to users' autonomy in online spaces (Kozyreva et al., 2020). Therefore, to use new technologies safely and for their own benefit, users of digital technologies need to acquire certain skills, such as problem-solving abilities and digital literacy (Livingstone & Helsper, 2010). Moreover, given the differences between online and in-person social interactions, socially conscious use of technology is essential (Kozyreva et al., 2020). In this sense, digital maturity shifts the current paradigm from focusing on issues related to technology use to exploring the beneficial applications of technology that go beyond a user's ability to operate mobile devices.

The concept of digital maturity incorporates aspects of previous approaches, such as the importance of digital literacy in the sense of digital competence, which focuses on practical knowledge and skills for using digital technologies. Existing approaches to digital literacy vary widely, and some definitions also include aspects of critical thinking or socio-emotional dimensions, such as knowing how to behave appropriately online (Audrin & Audrin, 2022; Eshet, 2004), some of which are reflected in digital maturity (e.g., showing respect towards others online). Other existing approaches recognize the importance of social responsibility (i.e., digital citizenship; Richardson & Milovidov, 2019) and the experience of autonomy in online contexts (Przybylski et al., 2010). The concept of digital maturity encompasses these different existing approaches and adds relevant new approaches (e.g., Individual Growth in Digital Contexts) based on its theoretical underpinnings. It goes beyond existing constructs by combining those aspects of digital technology use that are critical to both personal development and social adjustment. It provides a holistic and multidimensional approach that is currently lacking. Thus, digital maturity recognizes young people's ubiquitous use of digital technologies that are integrated into their daily lives and experiences.

16.2.2 The Multidimensionality of Digital Maturity

The derivation and validation of digital maturity as a multidimensional construct followed an extensive procedure, which is illustrated in Figure 16.1 (Laaber et al., 2023).

Based on this procedure, we (i) derived the multidimensional conceptualization of digital maturity (literature search, focus groups, and expert ranking), (ii) operationalized the dimensions and underlying measurement framework, (iii) validated the conceptualization of digital maturity and underlying measurement (Study 1 and 2; Laaber et al., 2023), and (iv) explored potential antecedents and outcomes (e.g., Koch et al., 2024; Laaber et al., 2024).

We view digital maturity as a multidimensional concept that is formed by its dimensions, and we identified three capacities and ten dimensions of digital maturity by applying ideas from psychosocial maturity and self-determination to the issues young people face in the digital world. Using a theory-driven approach, we developed a scale to measure this broad concept (see Laaber et al., 2023 for more details; see Figure 16.2).

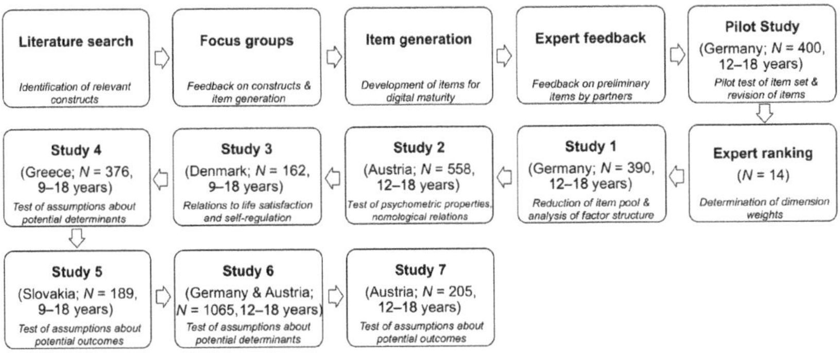

Figure 16.1 *Visualization of the procedure for the conceptualization and measurement taxonomy of digital maturity*

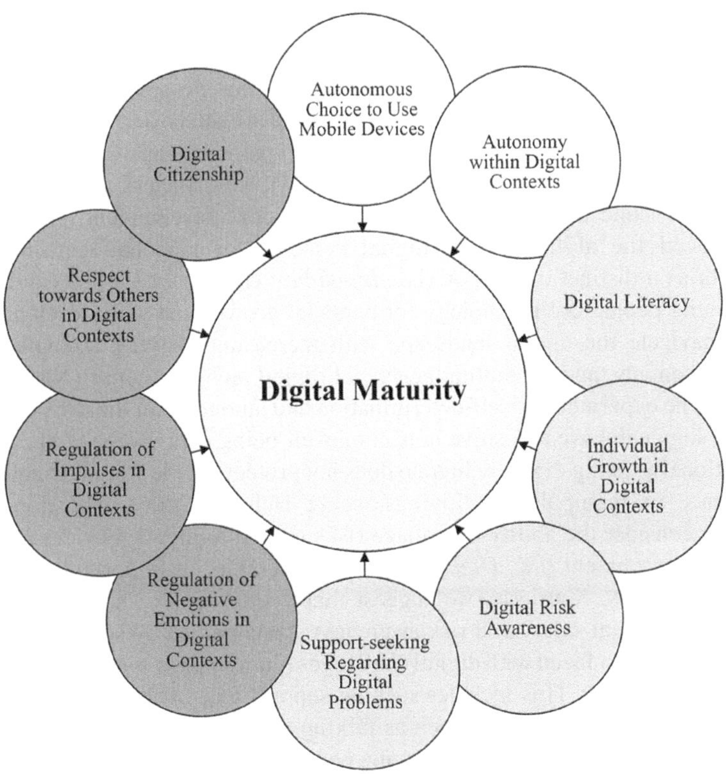

Figure 16.2 *The conceptualization of digital maturity*

First, the *capacity to use digital technologies in an autonomous and self-determined way* addresses the risks associated with the technical characteristics of digital technologies engineered to keep users' attention and influence their decision-making online, often for commercial interests (Kozyreva et al., 2020). It includes two dimensions: *Autonomous Choice to Use Mobile Devices* describes the degree to which young people use mobile devices out of their own choice rather than out of a sense of obligation or compulsion. This feeling of compulsion may be expressed, for example, in the urge to remain constantly connected and to go online frequently to avoid feeling left out of other people's online activities (Koessmeier & Büttner, 2021). The second aspect of autonomous decision-making is *Autonomy within Digital Contexts,* which refers to deliberately choosing and pursuing goals in the digital environment according to one's own interests and values. *Autonomy within Digital Contexts* extends beyond the autonomous decision to use mobile devices, encompassing the capacity to set and pursue goals online that align with one's personal interests and values (Laaber et al., 2023).

Second, the *capacity to master increasing digital challenges and solve problems* is related to the digital skills needed in rapidly changing technological environments (Laaber et al., 2023). We suggest that possessing fundamental technical skills (i.e., *Digital Literacy*) in order to safely and effectively use mobile devices and the internet is key to mastering increasing digital challenges and solving problems (Livingstone et al., 2021). Nevertheless, digital literacy is not the same as using technology for personal development. Hence, we introduced the ability to use digital contexts for personal learning and growth as a distinct dimension (i.e., *Individual Growth in Digital Contexts*). As young people use technology for personal growth and skill development, they navigate the digital landscape with increasing mastery, strengthening their autonomy, and promoting beneficial digital technology use (Nota et al., 2011). The experience of self-determination and autonomy in the use of digital technology can have a positive impact on well-being (Przybylski et al., 2010). Additionally, being digitally literate does not protect people from technology's influence or manipulation (Livingstone & Helsper, 2006). Therefore, we further consider the ability to manage risks related to mobile devices and the digital environment (i.e., *Digital Risk Awareness*) as an important aspect of digital maturity. Similarly, we suggest that young people should possess a sense of personal safety and risk awareness accompanied by support-seeking behaviours when faced with digital challenges (i.e., *Support-seeking Regarding Digital Problems*). This includes seeking support when encountering serious problems in digital contexts, such as talking to others about an issue (Appel et al., 2012). Considering that young people are often alone when facing challenging situations online, seeking help from parents, siblings, or friends is a valuable resource in such situations (Staksrud & Livingstone, 2009).

Third, the *capacity to interact adequately with others and contribute to society* combines aspects of self-regulation skills and attitudes towards others, which are relevant to the social context of digital technology use. For instance, regulating negative emotions when one is provoked by others online, such as through a critical comment, is more difficult than in in-person interactions (Wright, 2013). The use of emotion regulation strategies is crucial when young people engage with others in the digital environment, as coping with frustration can be particularly challenging due to the perceived psychological distance from others (Lipinski-Harten & Tafarodi, 2013; Wright, 2013). Therefore, digital maturity encompasses the ability to effectively control and regulate negative emotions (i.e., *Regulation of Negative Emotions in Digital Contexts*) and to regulate impulses occurring due to frustration in digital contexts (i.e., *Regulation of Impulses in Digital Contexts*). Additionally, we include respectful attitudes towards others in digital settings (i.e., *Respect towards Others in Digital Contexts*), as well as feeling responsibility towards society (i.e., *Digital Citizenship*) to cover positive social engagement online (Armstrong-Carter & Telzer, 2021). *Respect towards Others* includes aspects such as accepting different opinions online. Moreover, *Digital Citizenship* describes the use of digital contexts for contributions to society, for instance, by actively engaging in social, societal, and environmental causes. This can include involvement by young people in the enhancement of their communities and speaking up for important issues such as the protection of the environment (Laaber et al., 2023).

16.3 MEASUREMENT OF DIGITAL MATURITY

Digital maturity is conceptualized as a higher-order construct, which is constituted by its underlying dimensions (Bollen & Diamantopoulos, 2017). According to this structure, digital maturity is not conceived as a common factor that defines its components; rather, digital maturity is the sum of the relevant components. As outlined above, each dimension of digital maturity is a single latent variable captured by several reflective indicators of digital technology behaviours (Laaber et al., 2023).

To develop the measurement of digital maturity – the Digital Maturity Inventory (DIMI) – we conducted two studies. Following the typical scale development procedure, the aim of the first study ($N = 390$, German sample, 12–18 years) was to decrease the number of items from the original item pool and examine the factor structure of digital maturity. The digital maturity items were created by three experts and were both self-developed and adaptations of existing measures (see Laaber et al., 2023).

First, we established anchor items that reflected the digital maturity dimensions and chose items from the pool that correlated the highest with each anchor item, respectively. Second, we used exploratory factor analysis (EFA) to verify the operationalization based on the ten dimensions. This led to a selection of 40 DIMI items. The aim of the second study (*N* = 558, Austrian sample, 12–18 years) was to confirm the DIMI's initial factor structure and to investigate the nomological network in which the DIMI is embedded. We tested each dimension's factorial structure by applying confirmatory factor analysis (CFA). We removed items with low factor loadings and retained a final sample of 32 items for the final DIMI (see Figure 16.3 for the final items; see Laaber et al., 2023).

Overall, we developed a scale with good psychometric properties. As shown in subsequent sections, our concept and measure of digital maturity are integrated into a nomological network of personality development and self-regulation yet remain distinct from existing concepts and measures.

16.4 ANTECEDENTS AND OUTCOMES OF DIGITAL MATURITY

In a series of studies, we have established the importance of digital maturity as a novel concept and demonstrated that it is distinct from other related constructs and personality traits. For an overview of the study outcomes, refer to Figures 16.4 and 16.5. Prior to conducting the studies, we obtained ethical approval from the university's ethics committee and adhered to strict ethical guidelines for studies involving minors. This entailed obtaining informed consent from both the parent and the participating child at the onset of the study, as well as debriefing and final consent from both parties at the conclusion of the study.

In examining the *determinants of digital maturity*, we considered the influences of family socioeconomic status (SES) and parental support of their child's technology use (i.e., active parental mediation; Livingstone et al., 2011). Previous research has indicated that SES influences family technology use and youth online opportunities (Faltýnková et al., 2020; Sun et al., 2021). A study of a large sample (*N* = 1065) of German and Austrian adolescents between the ages of 12 and 18 found that adolescents from families with higher socioeconomic backgrounds exhibited greater digital maturity (Koch et al., 2024). It is noteworthy that this relationship was fully mediated by parents' active mediation styles. Parents with higher socioeconomic status engaged in more active mediation, or active involvement in their children's interactions with digital environments. This included activities such as explaining to their child how to use digital technologies, encouraging exploration, and providing guidance (Koch et al., 2024). In contrast, restrictive parenting practices, such as limiting digital activities or screen time, were not found to be associated with digital maturity (Koch et al., 2024).

Capacity	Dimensions	Items "When using a mobile device..."	Factor loadings
The Capacity to Use Digital Technologies in an Autonomous and Self-determined Way	1. Autonomous Choice to Use Mobile Devices	(α = .77 *CR* = .78 *AVE* = .55)	
		1. … I am online because I feel like I have to be online all the time.	.85
		2. … I am online because otherwise I feel like I am missing out on something.	.78
		3. … I have the feeling that it is controlling my life.	.57
	2. Autonomy Within Digital Contexts	(α = .70 *CR* = .70 *AVE* = .44)	
		1. … I choose the content I want to see.	.68
		2. … I do the things that I like.	.65
		3. … I decide what I do.	.65
The Capacity to Master Increasing Digital Challenges and Solve Problems	3. Digital Literacy	(α = .82 *CR* = .82 *AVE* = .61)	
		1. I know how to change the privacy settings (for example, deactivate cookies).	.83
		2. I know how to adjust the privacy settings of social media sites (for example, Instagram, Snapchat, TikTok).	.80
		3. I know how to store photos, documents or other files in the cloud (for example, Google Drive, iCloud).	.71
	4. Individual Growth in Digital Contexts	(α = .81 *CR* = .81 *AVE* = .59)	
		1. … I learn new things.	.79
		2. … I learn something useful.	.76
		3. … I learn new skills.	.76
	5. Digital Risk Awareness	(α = .85 *CR* = .85 *AVE* = .66)	
		1. … I am very careful.	.83
		2. … my own safety is very important to me.	.77
		3. … I make sure to be careful.	.84
	6. Support-Seeking regarding Digital Problems	(α = .80 *CR* = .81 *AVE* = .52)	
		1. … I ask others for help when I have a problem.	.80
		2. … and I don't know what to do, I ask others for help.	.77
		3. … and there is a technical problem, I ask others for help (for example, a friend, parent, sibling).	.69
		4. … and I have problems with others on the internet, I get help.	.61
The Capacity to Interact Adequately with Others and Contribute to Society	7. Regulation of Negative Emotions in Digital Contexts	(α = .84 *CR* = .85 *AVE* = .65)	
		1. … and I become annoyed or upset online, it takes me a long time to feel better.	.86
		2. … and I become annoyed or upset online, I stay in a bad mood for a long time.	.79
		3. … and I become annoyed or upset online, there are only few things that can make me feel better.	.76
	8. Regulation of Impulses in Digital Contexts	(α = .77 *CR* = .78 *AVE* = .54)	
		1. … and somebody criticizes me online or in a text, I immediately react without considering the consequences.	.81
		2. … and a message makes me angry, I react too quickly and then later regret the way I responded.	.71
		3. … and somebody insults me, I try to get back at them.	.68
	9. Respect Towards Others in Digital Contexts	(α = .83 *CR* = .83 *AVE* = .55)	
		1. … I watch my language when I disagree with someone, so that what I say doesn't come across as mean.	.76
		2. … I care about the feelings of others.	.79
		3. … I make sure that pictures I post or send of other people will not insult them or get them into trouble.	.70
		4. … I respect the opinions and knowledge of others.	.70
	10. Digital Citizenship	(α = .85 *CR* = .85 *AVE* = .66)	
		1. … I use it to improve life in my neighborhood, town or world.	.79
		2. … I use the internet to support campaigns for issues like environmental protection or to spread awareness for climate change.	.82
		3. … I use it to stand up for things that really matter in the world.	.82

Note: α = Cronbach's alpha, *CR* = Construct Reliability, *AVE* = Average Variance Extracted. *N* = 558.

Figure 16.3 *Conceptualization of digital maturity and psychometric properties of the final digital maturity inventory (DIMI)*

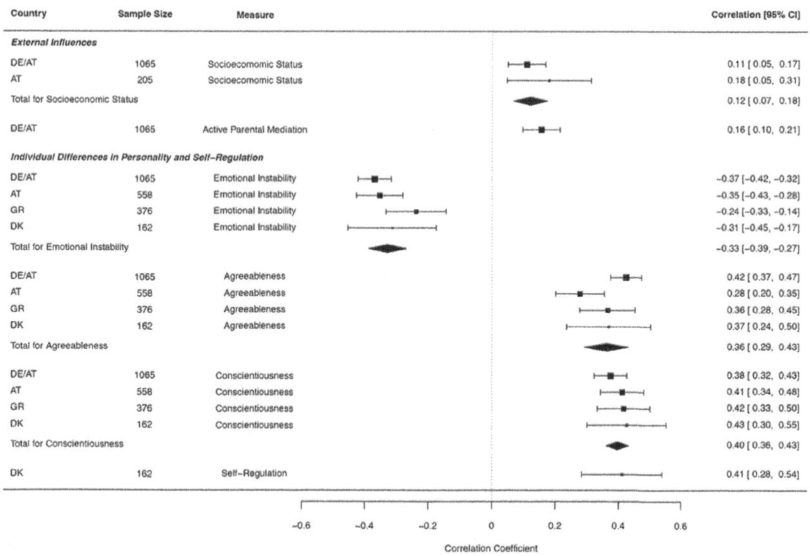

Note: DE = Germany. AT = Austria. GR = Greece. DK = Denmark. 95% CI = 95% Confidence Interval.

Figure 16.4 Determinants of digital maturity

The study also revealed a positive relationship between digital maturity and the personality traits of agreeableness and conscientiousness, and a negative relationship with emotional instability, as measured by scales of the Big Five personality domains (e.g., Soto & John, 2017). We tested the relationship between digital maturity and these personality traits in several studies because previous research has shown that as people move towards greater personality maturity, they exhibit higher levels of agreeableness and conscientiousness and lower levels of emotional instability (Bleidorn, 2015; Roberts & Mroczek, 2008).

In the aforementioned study with a sample of 558 teenagers in Austria (ages 12–18), we observed a moderately positive correlation between digital maturity and agreeableness and conscientiousness, and a negative correlation with emotional instability. These findings are consistent with the theoretical proposition that digital maturity is anchored in personality maturity. As anticipated, no significant relationship was found between extraversion and digital maturity. Interestingly, we found an unexpectedly strong positive relationship between open-mindedness and digital maturity. Regarding digital technology use, a stepwise hierarchical linear regression analysis demonstrated

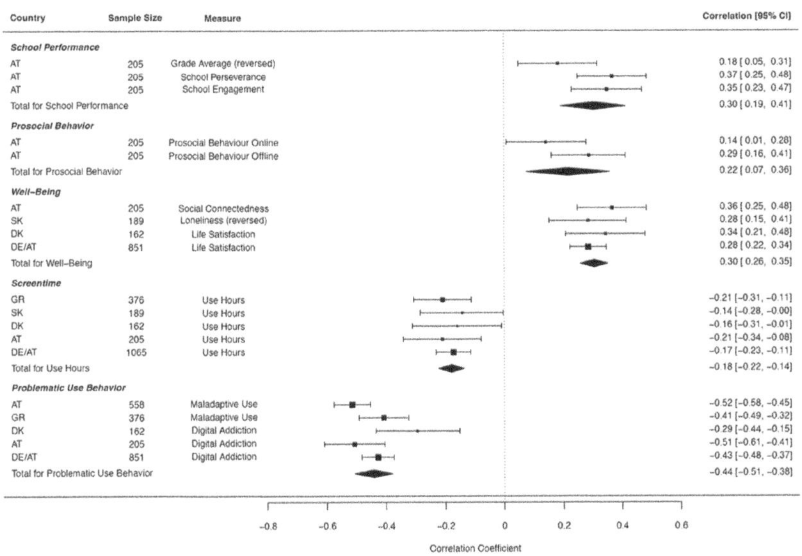

Note: AT = Austria. SK = Slovakia. DK = Denmark. DE = Germany. GR = Greece. 95% CI = 95% Confidence Interval. The grade average in Austria is scored from 1 (best grade) to 5 (fail). It was thus reversed to show that better grades correlate positively with digital maturity.

Figure 16.5 Outcomes of digital maturity

that digital maturity was a stronger predictor of problematic mobile device use than personality traits alone, even when personality traits were taken into account (Laaber et al., 2023).

We conducted two additional studies, one with a sample of 376 participants between the ages of 9 and 18 in Greece using the DIMI in two age versions, and the other with a sample of 162 participants between the ages of 12 and 18 in Denmark, which confirmed the associations between digital maturity and personality maturity. Data from several European countries provided further evidence that digital maturity, as measured by the DIMI, is rooted in personality maturity, as reflected in conscientiousness, agreeableness, and emotional stability.

We used the Danish study to further explore the relationship between digital maturity and general self-regulation, as previous research has shown that self-regulation and self-control are important skills for young people to use ICTs effectively (Reinecke et al., 2022; Schulz van Endert, 2021). The results confirmed that digital maturity is moderately positively related to self-regulation, but not to the extent that they are conceptually identical.

Regarding the *outcomes of digital maturity*, we have examined the relationships between digital maturity and relevant outcomes, including academic performance, prosocial behaviour, well-being, screen time, and problematic use behaviours, in several European studies.

In a study of Austrian adolescents aged 12–18 years ($N = 205$), we investigated several areas of personal growth, prosocial behaviour, and social relations that we added as new constructs to the established knowledge base. In terms of personal growth, several aspects of adolescents' academic performance were included. Looking at their personal growth and social relationships is highly relevant as the digital environment and young people's behaviour in it have become an integral part of their lives. Across all measures, adolescents who were more digitally mature also performed better in school. Digital maturity was positively associated with parent-reported grade average, school perseverance, and school engagement (Buerger et al., 2023). In terms of social aspects, digital maturity was positively associated with prosocial behaviour both online and offline. The sample further revealed a positive association between digital maturity and social connectedness, as measured using items from Kushlev et al. (2017), which is an important aspect of young people's social integration and well-being. Established correlations of digital maturity with higher socioeconomic status, lower hours of use, and lower digital addiction were also supported. Overall, the Austrian study provides further insights into the relationships between digital maturity and other relevant aspects of young people's lives.

Using a Slovak version of the DIMI, we examined the relationship between digital maturity and loneliness, and whether individuals are better able to cope with negative social experiences in digital contexts, such as cyber-ostracism, as their digital maturity increases. We translated the DIMI into Slovak and administered both age versions to a sample of 189 participants aged 9–11 and 12–18 in Slovakia. In addition to the DIMI, we also tested the effects of a cyber-ostracism manipulation on emotional responses and included a measure of general loneliness. We presented participants with a thought experiment in which they were asked to imagine a situation in which they were being excluded by others in an online context. After that, we assessed participants' emotional responses to the scenario at various time points. Consistent with previous studies, digital maturity was negatively correlated with hours of mobile device use and general loneliness. However, the DIMI did not correlate with responses to the cyber-ostracism scenario, suggesting that immediate responses to such situations may be negative in all children and that further research is needed in this area.

In another study, we invited participants aged 12–18 years who were part of the first study on digital maturity and SES in Germany and Austria ($N = 1065$) to a follow-up study approximately one month later ($N = 851$) to test several potential outcomes of digital maturity. Consistent with previous studies, the longitudinal results replicated that high digital maturity is positively related to life satisfaction. Our study in Denmark also found a significant positive correlation between digital maturity and general life satisfaction. In addition, a multiple regression analysis with digital maturity and self-regulation as predictors and life satisfaction as the dependent variable showed that only digital maturity was a significant predictor of life satisfaction. These findings suggest that digital maturity is related to life satisfaction and that this relationship cannot be explained by general self-regulatory competencies.

We also included a measure of the subjective total use of mobile devices (i.e., screen time) in most of the studies to investigate whether digital maturity is negatively related to total screen time. We found consistent negative associations between digital maturity and hours of use, suggesting that young people with higher levels of digital maturity tend to use their mobile ICTs less than those with lower levels of digital maturity. However, the results also show that digital maturity and screen time are two different concepts, supporting the idea that digital maturity focuses on how technology is used rather than how much (Laaber, 2024).

The studies in Austria, Germany, Greece, and Denmark were also used to explore the relationships between digital maturity and problematic device use, as we assumed that digital maturity should support young people in using technology in positive ways. In one of the studies in Austria ($N = 558$) and the study in Greece ($N = 376$), we measured problematic device use with four items that we adapted from two digital addiction scales (Andreassen et al., 2012; Hawi et al., 2019). In the studies in Denmark ($N = 162$), Austria ($N = 205$), and Austria and Germany ($N = 851$), we measured problematic use with the Digital Addiction Scale for Teenagers (DAST; Seema et al., 2022). We found consistently negative correlations between digital maturity and problematic use, suggesting that the higher the level of digital maturity, the less likely young people are to engage in problematic forms of device use.

16.5 DISCUSSION

The goal of our research was to develop and establish a conceptual framework for understanding children's and adolescents' digital maturity. We have based the development of our humanistic framework on theories of psychosocial development and self-determination theory (Laaber et al., 2023). We have created a novel and comprehensive definition of digital maturity and thoroughly explored potential determinants and outcomes. In our view, digital maturity is characterized by the autonomous use of digital technologies that promote psychological development and well-being while protecting against potential risks and taking into account societal demands. Studies of children and adolescents in several countries are consistent with our basic proposition that digital maturity is related to personality development and emerges through learning and socialization. In part, it reflects social inequalities in society. Most importantly, our findings indicate that digital maturity is linked to behaviours and experiences in the digital space and beyond.

Figures 16.4 and 16.5 provide a summary of the relationships between digital maturity and potential determinants (Figure 16.4) and outcomes (Figure 16.5). The results indicated that digital maturity is moderately related to personality traits that reflect personality maturity, such as conscientiousness, agreeableness, and emotional stability. However, digital maturity is not solely based on personality maturity. It extends beyond personality maturity and self-regulation skills in predicting outcomes such as maladaptive digital behaviour and life satisfaction.

We have developed a novel instrument and taxonomy to measure digital maturity that has been extensively validated (Laaber et al., 2023). In doing so, we have attempted to make digital maturity as measurable as possible with the DIMI. At the same time, we recognize that further development of both the construct and the attempts to measure digital maturity will greatly benefit the understanding of digital maturity.

16.5.1 Practical Implications

The DIMI will enable the provision of specific recommendations and insights for a range of stakeholders, including families, schools, and policymakers. Parents can utilize digital maturity insights in their parenting behaviours, such as actively supporting their child's exploration of digital opportunities while focusing on maintaining autonomy in digital contexts. Teachers may also benefit from the insights gained by incorporating the findings into their teaching of digital technology use. Indeed, as part of a related research project called "BeWEEN – Be well and green when digital", we have developed educational materials on digital maturity, which can be used by teachers. Policymakers may utilize the DIMI, which has already been translated into several languages, to monitor digital maturity across countries. We posit that monitoring digital maturity is beneficial for several reasons. First, the measurement of digital maturity provides relevant stakeholders and practitioners with essential information about the current situation and allows for comparisons between different social groups and countries. Second, our taxonomy can assist in tailoring interventions to enhance specific aspects of digital maturity. Third, measuring digital maturity allows for a systematic evaluation of interventions and can accompany policy implementation.

16.5.2 Avenues for Future Research

To gain further insights into the development of digital maturity, further research is required (e.g., Christensen et al., 2023). In particular, it is necessary to explore how digital maturity develops over time by applying longitudinal studies. Furthermore, intervention studies can assist in elucidating the causal influence of digital maturity on crucial outcome variables and the mechanisms that mediate the effects of digital maturity. Additionally, we consider it a valuable endeavour to investigate the measurement invariance of the construct across countries and the relevance of individual dimensions of digital maturity (e.g., Hofmans et al., 2024).

In conclusion, we have developed a novel theoretical concept of digital maturity and a reliable and valid measure of digital maturity, the Digital Maturity Inventory (DIMI), which has been tested in several studies across Europe. The results provided valuable insights into potential determinants of digital maturity, including environmental influences (e.g., SES), personality, and self-regulation, as well as potential outcomes of digital maturity, including school performance, prosocial behaviour, well-being, screen time, and problematic use.

AUTHORS' NOTE

The work presented here has been developed in the context of the Digymatex project. This project has received funding from the European Union's Horizon 2020 research and innovation programme under grant agreement No 870578.

REFERENCES

Andreassen, C. S., Torsheim, T., Brunborg, G. S., & Pallesen, S. (2012). Development of a Facebook addiction scale. *Psychological Reports*, *110*(2), 501–517. https://doi.org/10.2466/02.09.18.PR0.110.2.501-517.

Appel, M., Holtz, P., Stiglbauer, B., & Batinic, B. (2012). Parents as a resource: Communication quality affects the relationship between adolescents' internet use and loneliness. *Journal of Adolescence*, *35*(6), 1641–1648.

Armstrong-Carter, E., & Telzer, E. H. (2021). Advancing measurement and research on youths' prosocial behavior in the digital age. *Child Development Perspectives*, *15*(1), 31–36.

Audrin, C., & Audrin, B. (2022). Key factors in digital literacy in learning and education: A systematic literature review using text mining. *Education and Information Technologies*, *27*, 7395–7419. https://doi.org/10.1007/s10639-021-10832-5.

Bastiaensens, S., Pabian, S., Vandebosch, H., Poels, K., Van Cleemput, K., DeSmet, A., & De Bourdeaudhuij, I. (2016). From normative influence to social pressure: How relevant others affect whether bystanders join in cyberbullying: Normative influence on cyberbullying bystanders. *Social Development*, *25*(1), 193–211.

Bell, V., Bishop, D. V., & Przybylski, A. K. (2015). The debate over digital technology and young people. *BMJ*, *351*, h3064. doi:10.1136/bmj.h3064.

Bleidorn, W. (2015). What accounts for personality maturation in early adulthood? *Current Directions in Psychological Science*, *24*(3), 245–252.

Bollen, K. A., & Diamantopoulos, A. (2017). In defense of causal-formative indicators: A minority report. *Psychological Methods*, *22*(3), 581–596.

Bozzola, E., Spina, G., Agostiniani, R., Barni, S., Russo, R., Scarpato, E., Di Mauro, A., Di Stefano, A. V., Caruso, C., Corsello, G., & Staiano, A. (2022). The use of social media in children and adolescents: Scoping review on the potential risks. *International Journal of Environmental Research and Public Health*, *19*(16), 9960. https://doi.org/10.3390/ijerph19169960.

Brautsch, L. A., Lund, L., Andersen, M. M., Jennum, P. K., Folker, A. P., & Andersen, S. (2023). Digital media use and sleep in late adolescence and young adulthood: A systematic review. *Sleep Medicine Reviews*, *68*, 101742. https://doi.org/10.1016/j.smrv.2022.101742.

Buckingham, D. (2015). Defining digital literacy: What do young people need to know about digital media? *Nordic Journal of Digital Literacy*, *10*, 21–35.

Buerger, S., Holzer, J., Yanagida, T., Schober, B., & Spiel, C. (2023). Measuring adolescents' well-being in schools: The adaptation and translation of the EPOCH measure of adolescent well-being: A validation study. *School Mental Health*, *15*(2), 611–626. https://doi.org/10.1007/s12310-023-09574-1.

Christensen, R. N., Siddiqui, A., Valogianni, K., Florack, A., & Hubert, M. (2023). Towards a composite index for digital maturity: An unsupervised machine learning approach. The 15th Mediterranean Conference on Information Systems (MCIS) and the 6th Middle East & North Africa Conference on Digital Information Systems (MENACIS), 10903. https://easychair.org/publications/preprint/wvtx.

Deci, E. L., & Ryan, R. M. (2000). The "what" and "why" of goal pursuits: Human needs and the self-determination of behavior. *Psychological Inquiry*, *11*(4), 227–268.

Eshet, Y. (2004). Digital literacy: A conceptual framework for survival skills in the digital era. *Journal of Educational Multimedia and Hypermedia*, *13*(1), 93–106.

Faltýnková, A., Blinka, L., Ševčíková, A., & Husarova, D. (2020). The associations between family-related factors and excessive internet use in adolescents. *International Journal of Environmental Research and Public Health*, *17*(5), 1754. https://doi.org/10.3390/ijerph17051754.

Gan, B., Menkhoff, T., & Smith, R. (2015). Enhancing students' learning process through interactive digital media: New opportunities for collaborative learning. *Computers in Human Behavior*, *51*, 652–663. https://doi.org/10.1016/j.chb.2014.12.048.

Greenberger, E., Josselson, R., Knerr, C., & Knerr, B. (1975). The measurement and structure of psychosocial maturity. *Journal of Youth and Adolescence*, *4*(2), 127–143.

Haddon, L., & Livingstone, S. (2017). Risks, opportunities, and risky opportunities: How children make sense of the online environment. In F. C. Blumberg & P. J. Brooks (Eds), *Cognitive Development in Digital Contexts* (pp. 275–302). Elsevier.

Hawi, N. S., Samaha, M., & Griffiths, M. D. (2019). The digital addiction scale for children: Development and validation. *Cyberpsychology, Behavior, and Social Networking*, *22*(12), 771–778. https://doi.org/10.1089/cyber.2019.0132.

Helsper, E. J., Schneider, L. S., van Deursen, A. J. A. M., & van Laar, E. (2020). *The Youth Digital Skills Indicator: Report on the Conceptualization and Development of the ySKILLS Digital Skills Measure*. KU Leuven, Leuven: ySKILLS.

Hofmans, L., van der Stappen, A., & van den Bos, W. (2024). Developmental structure of digital maturity. *Computers in Human Behavior*, *157*, 108239. https://doi.org/10.1016/j.chb.2024.108239.

Huang, S., Lai, X., Li, Y., Luo, Y., & Wang, Y. (2021). Understanding juveniles' problematic smartphone use and related influencing factors: A network perspective. *Journal of Behavioral Addictions*, *10*(3), 811–826. https://doi.org/10.1556/2006.2021.00048.

Kardefelt Winther, D., Livingstone, S., & Saeed, M. (2019). *Growing Up in a Connected World* (Innocenti Research Report). UNICEF Office of Research, Innocenti. https://www.unicef-irc.org/publications/pdf/GKO%20Summary%20Report.pdf.

Kim, H., Schlicht, R., Schardt, M., & Florack, A. (2021). The contributions of social comparison to social network site addiction. *PloS One*, *16*(10), e0257795. https://doi.org/10.1371/journal.pone.0257795.

Koch, T., Laaber, F., & Florack, A. (2024). Socioeconomic status and young people's digital maturity: The role of parental mediation. *Computers in Human Behavior*, *154*, 108157. https://doi.org/10.1016/j.chb.2024.108157.

Koessmeier, C., & Büttner, O. B. (2021). Why are we distracted by social media? Distraction situations and strategies, reasons for distraction, and individual differences. *Frontiers in Psychology*, *12*, 711416.

Kozyreva, A., Lewandowsky, S., & Hertwig, R. (2020). Citizens versus the internet: Confronting digital challenges with cognitive tools. *Psychological Science in the Public Interest*, *21*(3), 103–156.

Kushlev, K., Proulx, J. D., & Dunn, E. W. (2017). Digitally connected, socially disconnected: The effects of relying on technology rather than other people. *Computers in Human Behavior*, *76*, 68–74. https://doi.org/10.1016/j.chb.2017.07.001.

Laaber, F. (2024). *A Psychological Perspective on Self-Determined Digital Technology Use: Conceptualizing, Measuring, and Understanding Young People's Digital Maturity* (unpublished doctoral thesis). University of Vienna.

Laaber, F., Florack, A., Koch, T., & Hubert, M. (2023). Digital maturity: Development and validation of the Digital Maturity Inventory (DIMI). *Computers in Human Behavior*, *143*, 107709. https://doi.org/10.1016/j.chb.2023.107709.

Laaber, F., Koch, T., Hubert, M., & Florack, A. (2024). Young People's digital maturity relates to different forms of well-being through basic psychological need satisfaction and frustration. *Computers in Human Behavior*, *152*, 108077. https://doi.org/10.1016/j.chb.2023.108077.

Lipinski-Harten, M., & Tafarodi, R. W. (2013). Attitude moderation: A comparison of online chat and face-to-face conversation. *Computers in Human Behavior*, *29*(6), 2490–2493.

Livingstone, S., & Bober, M. (2004). Taking up online opportunities? Children's uses of the internet for education, communication and participation. *E-Learning and Digital Media*, *1*(3), 395–419.

Livingstone, S., Haddon, L., Görzig, A., & Ólafsson, K. (2011). *Risks and Safety on the Internet: The Perspective of European Children: Full Findings and Policy*

Implications from the EU Kids Online Survey of 9–16 Year Olds and their Parents in 25 Countries (EU Kids Online, Deliverable D4). EU Kids Online Network.
Livingstone, S., & Helsper, E. J. (2006). Does advertising literacy mediate the effects of advertising on children? A critical examination of two linked research literatures in relation to obesity and food choice. *Journal of Communication*, *56*(3), 560–584.
Livingstone, S., & Helsper, E. (2010). Balancing opportunities and risks in teenagers' use of the internet: The role of online skills and internet self-efficacy. *New Media & Society*, *12*(2), 309–329.
Livingstone, S., Mascheroni, G., & Staksrud, E. (2018). European research on children's internet use: Accessing the past and anticipating the future. *New Media & Society*, *20*(3), 1103–1122.
Livingstone, S., Mascheroni, G., & Stoilova, M. (2021). The outcomes of gaining digital skills for young people's lives and well-being: A systematic evidence review. *New Media & Society*, 1–27.
Meshi, D., Tamir, D. I., & Heekeren, H. R. (2015). The emerging neuroscience of social media. *Trends in Cognitive Sciences*, *19*(12), 771–782. https://doi.org/10.1016/j.tics.2015.09.004.
Nota, L., Soresi, S., Ferrari, L., & Wehmeyer, M. L. (2011). A multivariate analysis of the self-determination of adolescents. *Journal of Happiness Studies*, *12*(2), 245–266.
Nyrhinen, J., Uusitalo, O., Frank, L., & Wilska, T. A. (2022). How is social capital formed across the digital-physical servicescape? *Digital Business*, *2*(2), 100047.
Ólafsson, K., Livingstone, S., & Haddon, L. (2013). *Children's Use of Online Technologies in Europe. A Review of the European Evidence Base*. LSE, London: EU Kids Online. http://eprints.lse.ac.uk/50228/1/__Libfile_repository_Content_Livingstone,%20S_Children%E2%80%99s%20use%20of%20online%20technologies%20in%20Europe(lsero).pdf.
Park, Y. J. (2013). Digital literacy and privacy behavior online. *Communication Research*, *40*(2), 215–236.
Paulich, K. N., Ross, J. M., Lessem, J. M., & Hewitt, J. K. (2021). Screen time and early adolescent mental health, academic, and social outcomes in 9- and 10-year-old children: Utilizing the Adolescent Brain Cognitive Development (ABCD) Study. *PloS One*, *16*(9), e0256591.
Przybylski, A. K., Rigby, C. S., & Ryan, R. M. (2010). A motivational model of video game engagement. *Review of General Psychology*, *14*(2), 154–166.
Reinecke, L., Gilbert, A., & Eden, A. (2022). Self-regulation as a key boundary condition in the relationship between social media use and well-being. *Current Opinion in Psychology*, *45*, 101296. https://doi.org/10.1016/j.copsyc.2021.12.008.
Richardson, J., & Milovidov, E. (2019). *Digital Citizenship Education Handbook: Being Online, Well-being Online, and Rights Online*. Council of Europe.
Roberts, B. W., & Mroczek, D. (2008). Personality trait change in adulthood. *Current Directions in Psychological Science*, *17*(1), 31–35. https://doi.org/10.1111/j.1467-8721.2008.00543.x.
Schulz van Endert, T. (2021). Addictive use of digital devices in young children: Associations with delay discounting, self-control and academic performance. *PloS One*, *16*(6), e0253058. https://doi.org/10.1371/journal.pone.0253058.
Scott, D. A., Valley, B., & Simecka, B. A. (2017). Mental health concerns in the digital age. *International Journal of Mental Health and Addiction*, *15*, 604–613. doi:10.1007/s11469-016-9684-0.
Seema, R., Heidmets, M., Konstabel, K., & Varik-Maasik, E. (2022). Development and validation of the Digital Addiction Scale for Teenagers (DAST). *Journal*

of Psychoeducational Assessment, *40*(2), 293–304. https://doi.org/10.1177/07342829211056З.
Shin, W., & Kang, H. (2016). Adolescents' privacy concerns and information disclosure online: The role of parents and the Internet. *Computers in Human Behavior*, *54*, 114–123. https://doi.org/10.1016/j.chb.2015.07.062.
Soto, C. J., & John, O. P. (2017). Short and extra-short forms of the Big Five Inventory–2: The BFI-2-S and BFI-2-XS. *Journal of Research in Personality*, *68*, 69–81. https://doi.org/10.1016/j.jrp.2017.02.004.
Staksrud, E., & Livingstone, S. (2009). Children and online risk: Powerless victims or resourceful participants? *Information, Communication & Society*, *12*(3), 364–387.
Staudinger, U. M., & Kunzmann, U. (2005). Positive adult personality development: Adjustment and/or growth? *European Psychologist*, *10*(4), 320–329.
Sun, X., Duan, C., Yao, L., Zhang, Y., Chinyani, T., & Niu, G. (2021). Socioeconomic status and social networking site addiction among children and adolescents: Examining the roles of parents' active mediation and ICT attitudes. *Computers & Education*, *173*, 104292. https://doi.org/10.1016/j.compedu.2021.104292.
Twenge, J. M., & Campbell, W. K. (2018). Associations between screen time and lower psychological well-being among children and adolescents: Evidence from a population-based study. *Preventive Medicine Reports*, *12*, 271–283. https://doi.org/10.1016/j.pmedr.2018.10.003.
Valkenburg, P. M., Meier, A., & Beyens, I. (2022). Social media use and its impact on adolescent mental health: An umbrella review of the evidence. *Current Opinion in Psychology*, *44*, 58–68. https://doi.org/10.1016/j.copsyc.2021.08.017.
Valkenburg, P. M., & Peter, J. (2009). Social consequences of the internet for adolescents: A decade of research. *Current Directions in Psychological Science*, *18*(1), 1–5.
Valkenburg, P. M., & Peter, J. (2011). Online communication among adolescents: An integrated model of its attraction, opportunities, and risks. *Journal of Adolescent Health*, *48*(2), 121–127. https://doi.org/10.1016/j.jadohealth.2010.08.020.
Vogel, E. A., Rose, J. P., Okdie, B. M., Eckles, K., & Franz, B. (2015). Who compares and despairs? The effect of social comparison orientation on social media use and its outcomes. *Personality and Individual Differences*, *86*, 249–256. https://doi.org/10.1016/j.paid.2015.06.026.
Wright, M. F. (2013). The relationship between young adults' beliefs about anonymity and subsequent cyber aggression. *Cyberpsychology, Behavior, and Social Networking*, *16*(12), 858–862.

17. Towards the digital future of AI generations: a conclusion

Terhi-Anna Wilska and Jussi Nyrhinen

The aim of this book was to construct a broad comprehension of today's young people as actors in the many spheres of a digitalized world. Drawing upon existing research and theories related to the agency of young people in digital environments, this book is among the first to construct a coherent narrative that envisages how communication, economy, work, and leisure intersect within the agency of young individuals across various digital landscapes. This book also focused on the underlying reasons behind these dynamics. The book highlighted novel findings and research perspectives while also examining the implications, and proposes recommendations to enhance young people's agency and competencies in digital environments. The chapters illustrated the various ways in which the digitalization of societies has changed young people's interaction with their families and peers and interpretation of information from different sources. The chapters also described how young people's digital agency is intricately linked to broader economic, digital, and social transformations, both in global and local contexts. Moreover, the rapid development of AI and how it is changing the roles of various actors in digital environments have been analysed in many chapters. Public and private digital platforms in work, leisure, education, and communication were revealed to expose their users to targeted influence but also to promote their digital agency.

Based on the findings of this book, it is evident that for the "diginative" generations (Kirschner & De Bruyckere, 2017; Prensky, 2001), digital technologies are double-edged or even triple-edged swords, containing severe risks, great opportunities, and useful tools for treating the social and mental problems that young people can experience in online and offline environments (Hollis et al., 2020; Stavropoulos et al., 2021). Recent research suggests that the use of the term "diginatives" for young people and "digital immigrants" for other age groups should be reconsidered. First, the number of age cohorts born into a digitalized world is increasing. Second, older generations, too, have adapted to digital devices and use social media and digital services fluently (Taipale et al., 2017); thus, they can hardly be called digital immigrants anymore. As noted in Chapter 5, young people are traditionally assumed to be active agents in digital communication with older generations. Yet, the chapter reveals that digital communication within the family is bidirectional, depending on the reasons and motivations for communication, as well as on digital channels, devices, and resources. Furthermore, new digital applications may restrain agency since the devices are also used for surveillance and control between family members. As contended in Chapter 6, this can be seen as protection of young people, but it may result in excessive control, and limited autonomy and agency.

Digital surveillance and control also operate in commercial digital landscapes. The rapid expansion of generative AI-based technology and the development of algorithmic surveillance and commercial targeting of internet users have made the youngest generations, known as Generations Z and Alpha (Chan & Lee, 2023; McCrindle, 2021), the main objects of these processes. However, as noted in Chapter 3, AI can increase young people's agency by providing them with relevant information and recommendations for products and services. Yet AI also constrains young people's agency, particularly for those who rely heavily on AI recommendations, by creating feedback loops that reinforce biased suggestions based on an individual's online behaviour. As AI-based algorithmic targeting of information concerns all digital environments, the credibility and transparency of information are issues that need increasing attention. The study presented in Chapter 4 recognized both careful and carefree attitudes towards non-transparent information obtained from social media. While most young people are cautious about non-transparency on social media, a small minority remains unconcerned. The same applies to the journalistic information of digital legacy media. Chapter 2 pointed out that digital and social media have blurred the lines between journalistic and other information. The authors also noted that digital environments, particularly social media, can make judging the credibility of information more difficult than before, which calls for more news media literacy skills.

However, young people's commercial privacy cannot primarily be attributed to knowledge gaps. Businesses that operate online want users to spend more time on their services and create hooks to intentionally draw users to buy a product or view an advertisement (Widdicks et al., 2022). Thus, there has also arisen a need for new kinds of digital consumer skills. Chapter 10 suggested that to protect young people's commercial privacy, we must not only increase their media literacy but also address the design of digital spaces. These environments should prioritize privacy protection and make it more convenient, avoiding *dark patterns* that encourage data sharing and excessive spending (Mathur et al., 2021). In addition to the advertising industry, various interest organizations and market regulators that target consumers struggle to find alternatives to profiling based on personal data and behaviour, particularly with regard to young people, but that is challenging, as contended in Chapter 9.

AI-based profiling and commercial influence have thus created new risks; therefore, it is crucial to improve the financial literacy of young people. It is important to understand the predictors of financial literacy and to detect the best practices and sources of financial education. By developing young people's digital financial competencies, they can make informed financial decisions and protect themselves from risks. The authors of Chapter 7 emphasize the importance of integrating financial education into school curricula, and empowering adolescents to handle financial matters autonomously. Furthermore, Chapter 8 highlights the special role of social media in young people's financial well-being. While family support is crucial for successful money management, peer support in social media groups is essential, especially during social and economic shocks, such as the COVID-19 pandemic.

The pandemic significantly increased the use of digital platforms in all age groups, both in work and leisure. As consumers needed home deliveries and private transport services, applications such as DoorDash, Fodora, Uber, and Bolt extended an economy in which individuals performed gig work via digital platforms. Chapter 12 argued that although this is a flexible way of working, considering recent economic turbulence, this development poses new risks to groups of people, often young migrants, who are already financially vulnerable. During the COVID-19 pandemic, gaming and gambling increased, particularly among young people (Wilska et al., 2021).

Chapter 11 discussed how young people's leisure within a digital space was orientated around consumption as a form of play. Although monetizing young people's emotions is an issue, it is difficult to prevent them from playing. Spending money within digital games is not a new phenomenon, but elements of gambling in games, such as loot boxes, are more recent phenomena (Sirola et al., 2023). This *gamblification* of digital games (Brock & Johnson, 2021) is connected to addictive gaming and gambling and shopping, as noted in Chapter 13. Personal factors, such as problems with self-control and self-regulation, make young people particularly vulnerable to impulsive and addictive behaviour on digital platforms. Feelings of loneliness and exclusion combined with online activity also expose individuals to other risks, such as the contents of the dark web (Sirola et al., 2022) and different kinds of digital crimes (Paat & Markham, 2020).

As mentioned above, digital platforms facilitate the data collection and profiling of individuals in different settings. In addition to issues related to privacy and commercial targeting, AI-based profiling may magnify inequalities and increase precarization and exclusion in societies. Chapter 14 explained how the use of AI-based technologies increased in the evaluation systems of education during the COVID-19 pandemic. According to the research, the use of algorithms in exam-grade standardization and online proctoring may increase inequalities; thus, they are particularly unfair for young people from disadvantaged backgrounds.

Too hasty and uncritical use of AI-based technologies may be problematic, but digital tools also have great potential for better inclusion of young people from disadvantaged backgrounds. Focusing on young refugees in Europe, Chapter 15 concluded that digital devices and digital skills are essential for young migrants, especially for managing the various risks and complexities of their lives, such as social marginality and lack of resources. As noted above, the triple-edged sword of digital technologies also has great potential for improving young people's well-being and resilience. Therefore, we must have relevant concepts to measure the effects of digitalization on young people and their capacity to cope in digital environments. Chapter 16 introduced a novel concept of *digital maturity* (i.e., skills and mindsets required to use digital technologies to enhance social integration, well-being, and academic performance). The chapter also presented a reliable and valid measure of digital maturity.

To conclude, in today's digitalized societies, young people are armed with great opportunities but also face severe risks. As the development of AI-based technologies is fast, it is necessary to look ahead to the future and detect new digital vulnerabilities that hinder young people's agency and well-being. In the future, an increasing number of different entities will attempt to influence young people's emotions, attitudes, and behaviours. Several digital platforms allow for information manipulation in novel ways that may harm society beyond individuals; thus, AI-generated misinformation and disinformation are seen as today's most severe global risks (World Economic Forum, 2024).

On the other hand, besides vulnerabilities, AI has enabled new kinds of active agency. Since organizations track citizens' behaviour in digital environments, and AI-based systems (e.g., decision support and targeted advertising) draw information from the online behaviour of people who are interacting with them, individuals intertwine with organizational decision-making through their digital footprint and direct recommendation systems. Thus, young people can practise agency in an automated manner with the data they produce. Prior studies have suggested that these shifts in agency may transfer power in organizational decision-making to consumers (e.g., Paavola & Cuthbertson, 2021).

Although AI-based systems provide opportunities, it can be argued that deepening commercialization is probably the most distinctive aspect of contemporary youth experiences in the digital world. Consequently, it is imperative that young people acquire enhanced media literacy, digital literacy, and financial literacy to effectively navigate the challenges of everyday life and fully capitalize on the opportunities afforded by digital environments. However, since AI is changing the digital landscape, particularly the commercial environments, traditional aids, such as media literacy programmes or knowledge- and attitude-based consumer education, may soon become outdated. What is thus needed in the future is better *AI literacy*. To promote this as well as the other necessary digital skills and literacies of the young generations, which could be named the *AI generations*, cooperation between young people, families, schools, universities, and public authorities is essential. It is important to recognize these issues and develop practical tools for advanced interventions that are strongly based on scientific evidence. We believe that the research presented in our book serves this purpose well.

REFERENCES

Brock, T., & Johnson, M. (2021). The gamblification of digital games. *Journal of Consumer Culture*, *21*(1), 3–13. https://doi.org/10.1177/1469540521993904.

Chan, C. K. Y., & Lee, K. K. (2023). The AI generation gap: Are Gen Z students more interested in adopting generative AI such as ChatGPT in teaching and learning than their Gen X and millennial generation teachers? *Smart Learning Environments*, *10*(1), 60.

Hollis, C., Livingstone, S., & Sonuga-Barke, E. (2020). The role of digital technology in children and young people's mental health – a triple-edged sword? *Journal of Child Psychology and Psychiatry*, *61*(8), 837–841.

Kirschner, P. A., & De Bruyckere, P. (2017). The myths of the digital native and the multitasker. *Teaching and Teacher Education*, *67*, 135–142.

Mathur, A., Kshirsagar, M., & Mayer, J. (2021). What makes a dark pattern… dark?: Design attributes, normative considerations, and measurement methods. *Proceedings of the 2021 CHI Conference on Human Factors in Computing Systems*, 1–18. https://doi.org/10.1145/3411764.3445610.

McCrindle, M. (2021). *Generation alpha*. Hachette.

Paat, Y. F., & Markham, C. (2020). Digital crime, trauma, and abuse: Internet safety and cyber risks for adolescents and emerging adults in the 21st century. *Social Work in Mental Health*, *19*(1), 18–40. https://doi.org/10.1080/15332985.2020.1845281.

Paavola, L., & Cuthbertson, R. (2021). Algorithms creating paradoxes of power: Explore, exploit, embed, embalm. *Information Systems Management*, *38*(4), 358–371.

Prensky, M. (2001). Digital natives, digital immigrants part 2: Do they really think differently? *On the Horizon*, *9*(6), 1–6. https://doi.org/10.1108/10748120110424843.

Sirola, A., Nyrhinen, J., Nuckols, J., & Wilska, T. A. (2023). Loot box purchasing and indebtedness: The role of psychosocial factors and problem gambling. *Addictive Behaviors Reports*, *18*, 100516. https://doi.org/10.1016/j.abrep.2023.100516.

Sirola, A., Nuckols, J., Nyrhinen, J., & Wilska, T. A. (2022). The use of the Dark Web as a COVID-19 information source: A three-country study. *Technology in Society*, *70*, 102012.

Stavropoulos, V., Motti-Stefanidi, F., & Griffiths, M. D. (2021). Risks and opportunities for youth in the digital era. *European Psychologist*, *27*(2), 86–101.

Taipale, S., Wilska, T. A., & Gilleard, C. J. (Eds). (2017). *Digital technologies and generational identity*. Routledge.

Widdicks, K., Remy, C., Bates, O., Friday, A., & Hazas, M. (2022). Escaping unsustainable digital interactions: Toward "more meaningful" and "moderate" online experiences. *International Journal of Human–Computer Studies*, 165, 102853.

Wilska, T. A., Sirola, A., Nuckols, J., & Nyrhinen, J. (2021). *A year of COVID-19 in three countries – A study on the effects of the COVID-19 pandemic on everyday life, consumption and digital behaviour in Finland, Sweden, and Great Britain*. YFI julkaisuja.9. Jyväskylän yliopisto. http://urn.fi/URN:ISBN:978-951-39-8829-6.

World Economic Forum. (2024). *The global risks report 2024*, 19th edition. World Economic Forum (weforum.org) (accessed 15 January 2024).

Index